The Palgrave Lacan Series

Series editor
Calum Neill
School of Psychology and Sociology
Edinburgh Napier University
Edinburgh, United Kingdom

Derek Hook
Duquesne University
Pittsburgh, USA

'Stanley Kubrick's *2001: A Space Odyssey* is the iconic soundtrack and cosmic screenplay of the contemporary world in every imaginary dimensionality. Beginning with but also quite apart from Slavoj Žižek, Daniel Bristow's masterful Lacanian coding unlocks *2001* in his book: offering the reader a tour, complete with color and sound, of the filmic imaginary, symbolic, and Real, a well-written and richly rewarding study.'

– Babette Babich, Executive Editor, *New Nietzsche Studies*

'In many ways the book reminds me of the first time watching the film. Within its pages are fascinating insights into not just the theory and writing of Lacan, but also new revelations regarding Kubrick's masterpiece. It, like the film to which it responds, will be a work to return to, in the firm expectation of gaining something new from it, with each visit.'

– Jon Greenaway, Literary Theorist and Blogger (@thelitcritguy)

'This amazing book jump cuts from Lacan to *2001* and back again. Bristow cuts into the Symbolic stuff of *A Space Odyssey* to show us that it is psychoanalysis that is at stake as the architecture of fantasy today, he cuts through Imaginary representations of Kubrick's classic that hypnotically bind us to it as if it were merely a sci-fi film, and he cuts us into a view of reality—of outer space and our inner space—that enables us to appreciate the place of the Real. This book proves that film is nothing without the traces of fantasy, and that Lacanian psychoanalysis is now secreted in every nook and cranny of cinematic enjoyment.'

– Ian Parker, Psychoanalyst, Manchester Psychoanalytic Matrix

Jacques Lacan is one of the most important and influential thinkers of the 20th century. The reach of this influence continues to grow as we settle into the 21st century, the resonance of Lacan's thought arguably only beginning now to be properly felt, both in terms of its application to clinical matters and in its application to a range of human activities and interests. The Palgrave Lacan Series is a book series for the best new writing in the Lacanian field, giving voice to the leading writers of a new generation of Lacanian thought. The series will comprise original monographs and thematic, multi-authored collections. The books in the series will explore aspects of Lacan's theory from new perspectives and with original insights. There will be books focused on particular areas of or issues in clinical work. There will be books focused on applying Lacanian theory to areas and issues beyond the clinic, to matters of society, politics, the arts and culture. Each book, whatever its particular concern, will work to expand our understanding of Lacan's theory and its value in the 21st century.

More information about this series at
http://www.palgrave.com/gp/series/15116

Daniel Bristow

2001: A Space Odyssey and Lacanian Psychoanalytic Theory

palgrave
macmillan

Daniel Bristow
Andover, UK

The Palgrave Lacan Series
ISBN 978-3-319-69443-6 ISBN 978-3-319-69444-3 (eBook)
https://doi.org/10.1007/978-3-319-69444-3

Library of Congress Control Number: 2017960753

Printed on acid-free paper

This Palgrave Macmillan imprint is published by Springer Nature
The registered company is Springer International Publishing AG
The registered company address is: Gewerbestrasse 11, 6330 Cham, Switzerland

For Louise,
Sky, Arlo, and Flora

Contents

List of Figures

CHAPTER 1

Overture

Abstract Following the structure of Stanley Kubrick's *2001: A Space Odyssey* (1968), this book opens with an 'Overture', showcasing the major themes of the work. Responding to the theses and theorems concerning cuts that Jean-Claude Milner extrapolates from the work of the Alexandres Koyré and Kojève, the chapter puts forth criteria for defining cuts as major or minor. *2001* of course begins with one of the most famous cuts in motion picture history; that from a bone thrown by an ape to a spacecraft in Earth's orbit four million years later. However, the argument is made that this cut is in fact minor in comparison to the major epistemological cut that inaugurates subjectivity itself, and which came earlier, with the first appearance of the astonishing solid ebon oblong, the monolith.

Keywords Overture • Cut • Monolith • Jacques Lacan • Jean-Claude Milner • Jacques-Alain Miller

'*A major cut separates the system of thought of Antiquity from the modern system of thought.*'[1] This is how Jean-Claude Milner summarises the central epistemological thesis of the great Hegelian philosopher Alexandre Kojève, a fundamental influence on the thinking of the French psychoanalytic successor to Sigmund Freud, Jacques Lacan. In Stanley Kubrick's collaborative film adaptation of Arthur C. Clarke's sci-fi novel(isation) *2001: A*

D. Bristow, *2001: A Space Odyssey and Lacanian Psychoanalytic Theory*, The Palgrave Lacan Series,
https://doi.org/10.1007/978-3-319-69444-3_1

Space Odyssey (1968) such a cut is ostensibly portrayed; that is, in the famous jump from a bone lobbed into the air by an angry ape on a desert Earth to an unidentified spacecraft floating amidst the dark of space just off Earth, an edit we might construe as a cut *from* the quick, if 'to cut to the quick' literally means to slice through flesh to the bone.[2] Yet, whilst separating them, this cut of course ties the two scenes together, the first being linked to the second through the filmic consuetude of an image-match edit—the bone falls alike to how the spacecraft is seen floating; their shapes are similar—and what is elided in the cut nonetheless suggests some continuity between the planet of the apes and the space conquest of the humans. It is as if—to combine G. W. F. Hegel and Martin Heidegger—here Spirit has derived from bone, in its thrownness.[3] However, the impression that one has led to the other is not given through an explicit statement (and therefore does not necessarily betray what Milner extrapolates from Kojève as his theorem, that '*there is never any synonymy between a notion belonging to Antiquity and a modern notion*'[4]): the cut is ambiguous, something is *subtracted* in it, perhaps similarly to the manner in which—in the Lacanian psychoanalytic conceptualisation—one's very own subjectivity is premised on its own subtraction, or *cut*; here in the subjective, rather than epistemological, definition.

As the popular Slovenian philosopher and Lacanian psychoanalytic theorist Slavoj Žižek might say, the barred subject—that is, in Lacan's notation: '$\$$'; the subject (S) cut by a rivening bar: '/'—is unseeable to itself, because *it is this cut* itself: its being cut out (of its reality) enframes it, constituting its reality. It is thus barred from being completely homogeneous with its reality, as its subtraction is *constitutive* of its reality. This is where Lacan's distinction between 'reality' and 'the Real' comes in: if the subject were to be homogeneous with(in) 'reality'—that is, due to the lack of subtraction productive of distinction or differentiation—this would thus entail the inexistence of both subject and reality *as such*; there would only be a homogeneous Real, indistinguishable as any element unto itself, as it is (subtracted) subjectivity that is required to produce any distinguishability whatsoever.[5] In other words, the subject cannot see itself seeing; it necessitously cannot see itself, in the third-person (dreams aside). Such an impossibility can only ever be anxiety-provoking fantasy, such as in the 'impossible sight that threatens you, of your own eyes lying there on the ground', as Lacan puts it, in relation to Oedipus unable to see his denucleated eyeballs, rent from their sockets by his own hand.[6] (The 'beyond of the infinite' that Dave Bowman (Keir Dullea) experiences during the Stargate sequence in the closing part of *2001* perhaps reaches this moment of anxiety of *seeing*; first, through to unconscious structuration itself, as if he were melding with it, after which is a series of cuts to and from the subject seeing himself in the film's last scenes).

However, to launch from this famous cut in *2001*, as is so often done, is to ignore the integrality of the monolith that precedes it. (In a similar vein to this appurtenance, Lacan mentions the huge stingray-like creature with abyssal black eyes caught at the end of Federico Fellini's *La Dolce Vita* (1960).[7]) A dark precursor if ever there was one; the monolith is a rectangular block as black as blackest night, unreflecting and impermeable. It has a likeness to a blank cinema or TV screen—like that over which the music of *2001*'s overture is played at its very beginning—yet it is precisely not a screen, due to its homogeny; for a screen separates, and through separating links, whereas the monolith is a Thing that only intrudes, interrupting previous continua and coordinates. It is in fact here that the true cut lies.

As an element of the Real, the only way to conceptualise its unrepresentability is through abstract aporias: it is like an impossibly complete set (in set theory), or an equally impossible reality without a lack, without the excerption-inclusion of a subject, or of object *a* (Lacan's always-already-lacking object), which—as Jacques-Alain Miller has stated—the subject becomes strictly equivalent with, *qua* subtracted, *qua cut*. This is an equivalency disavowed by the subject in their fundamental fantasy—that is, they do not see themselves as barred (i.e., $\$$ sees itself as S)—hence Lacan's formula for fantasy: '$\$ \diamond a$' (the barred subject and *a* separated by the multiform middle symbol, an enigmatic lozenge, which here can denote equivalency and disavowal at once).[8] Thus, to use Gottlieb Leibniz's famous word, the Thing is 'windowless', whereas—as Miller illustrates—a window is needed to get to anything like the soul (Fig. 1.1):

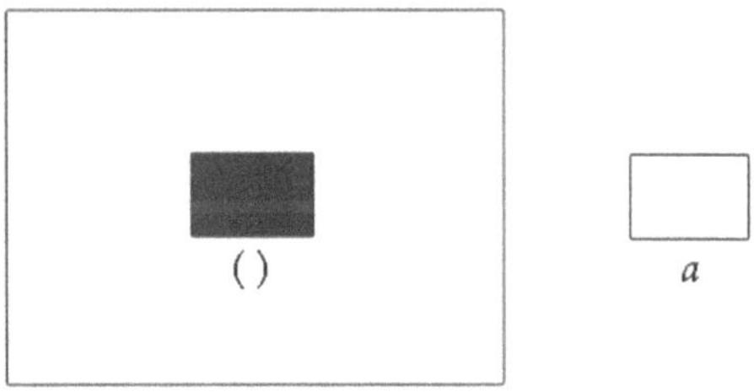

Fig. 1.1 Frame and Fragment, based on Miller's original diagram. Source: the author

> It is precisely *because* the object *a* is removed from the field of reality that it frames it. If I withdraw from the surface of this picture the piece I represented by a shaded square, I get what we might call a frame: a frame for a hole, but also a frame of the rest of the surface. Such a frame could be created by any window. So object *a* is such a surface fragment, and it is its subtraction from reality that frames it. The subject, as barred subject—as want-of-being—is this hole. As being, it is nothing but the subtracted bit. Whence the equivalency of the subject and object *a*.[9]

The monolith knows no such subtraction or withdrawal, but as *the Thing* it begins things: it does not screen, it does not frame, and most of all, *it does not deceive*, which is Lacan's definition for the cause of anxiety. 'Anxiety', Lacan states in his tenth Seminar; 'not only is it not without object, but it very likely designates the most, as it were, profound object, the ultimate object, the Thing. It's in this sense, as I've taught you to say, that anxiety is that which doesn't deceive.'[10] The apes in the opening segment of *2001*, 'The Dawn of Man', are certainly provoked by the monolith into something like an anxious frenzy; through this they stand upright and fashion a first tool, the bone as a weapon with which to kill. The docility prior to this provocation thus might be conceived as an entirely distinct *epistēmē* to this early science, from which it is thus separated by a major cut. This thus goes to Milner's summarisation of another great philosopher, lecturing at around the same time as Kojève, and as influential on Lacan, Alexandre Koyré, whose thesis is 'much more specific', in that it 'deal[s] with the status of science: *Modern science is entirely distinct from the* epistēmē *of Antiquity*.'[11] Thus, the true epistemological cut comes earlier than the famous filmic/temporal cut, which is in fact the film's second; the first cut is instantiated by the Thing, by confrontation with the monolith: it is the cut of subjectivity, that which instantiates subjectivity in the apes. The apes here seem to constitute themselves and their social link in relation to the monolith; they are thus constituted from without and made self-aware—brought into that age-old problematic philosophical state of *consciousness*—and it is this inaugural scene that fantasmatically highlights and links a great many of Lacan's subjectively constitutive formations of the unconscious: the mirror stage's recognition of fragmentarily unified bodies; the initially-supposed relation to the towering Other; existence as 'ex-sistence', not conceived as a purely inner essence, but more like something exoskeletal; and its related 'extimacy' (the unconscious as located and coming from 'outside', as 'the Other's discourse'). Although there are other cuts in *2001*, this is thus the most radical of them. More primordial than the others is the cut of subjectivity itself.

For Milner, Lacan's three theses are:

1. *There are major cuts.*
2. *Koyré's thesis is but a special case of Kojève's thesis; in other words, modern science is a major cut in Kojève's sense.*
3. *Psychoanalysis is intrinsically modern.*[12]

It is through interaction with *2001: A Space Odyssey* that we shall try to arrive, through its cuts, at these precise theses.

In his seventh Seminar Lacan explains Jeremy Bentham's usage of the English word 'fictitious' in his work *The Theory of Fictions*, claiming that this word has the same meaning as the French '*fictif*', but with the caveat that this is only so 'in the sense that every truth has the structure of fiction.'[13] It is this that will guide this study in Lacanian psychoanalysis, an intrinsically modern science, in its conjunction with an intrinsically modern science fiction; which is to say, in the meeting of two truths with the structure of fiction.[14]

Notes

1. Jean-Claude Milner, 'Lacan and the Ideal of Science', in *Lacan and the Human Sciences*, ed. by Alexandre Leupin (Lincoln: University of Nebraska Press, 1991) p.28.
2. Of *2001* as between novel and novelisation, Clarke wrote: 'but why write a novel, you may well ask, when we were aiming to make a movie? It's true that novelizations (ugh) are all too often produced afterward; in this case, Stanley had excellent reasons for reversing this process.' See Arthur C. Clarke, 'Foreword to the Millennial Edition', in *2001: A space Odyssey,* etc. [1968] (London: Orbit, 2001) p.x. Of its co-writing and accreditation, Clarke later explained: 'I felt that when the novel finally appeared it should be "by Arthur Clarke and Stanley Kubrick; based on the screenplay by Stanley Kubrick and Arthur Clarke"—whereas the movie should have the credits reversed. This still seems the nearest approximation to the complicated truth.' See Arthur C. Clarke, *The Lost Worlds of 2001* (London: Sidgwick and Jackson, 1972) p.31. I here follow the convention of the most recent editions, which give Clarke's name only.
3. See the comment '*Spirit is a bone*' in G. W. F. Hegel, *Phenomenology of Spirit* [1807], trans. by A.V. Miller (Oxford: Oxford University Press, 1977) p.208. This will be discussed at far greater length in the following

chapter, and beyond. Heidegger states in *Being and Time*: 'this characteristic of Dasein's being—this 'that it is'—is veiled in its "whence" and "whither", yet disclosed in itself all the more unveiledly; we call it the "*thrownness*" of this entity into its "there"; indeed, it is thrown in such a way that, as Being-in-the-world, it is the "there". The expression "thrownness" is meant to suggest the *facticity of its being delivered over*.' See Martin Heidegger, *Being and Time* [1927], trans. by John Macquarrie and Edward Robinson (Oxford: Blackwell, 1962) p.174.

4. Milner, 'Lacan and the Ideal of Science', in *Lacan and the Human Sciences*, p.28.
5. On the distinction between reality and Real, see the arguments throughout the course of Lacan's twentieth Seminar, in which reality is so often demonstrated to be phantasmatic. Specifically in reference to the highlighted dichotomy, see Jacques Lacan, *The Seminar of Jacques Lacan, Book XX: Encore: On Feminine Sexuality, the Limits of Love and Knowledge, 1972–1973*, ed. by Jacques-Alain Miller, trans. by Bruce Fink (New York: W. W. Norton, 1998) p.131. Thus, we could pictorialise this as:

With the cut:	Without the cut:
REAL	REAL
reality | real	
(cut: Differentiation)	(no Differentiation)

   In the Real alone there is thus no unification just as there is no separation.
6. Jacques Lacan, *The Seminar of Jacques Lacan, Book X: Anxiety* [1962–1963], ed. by Jacques-Alain Miller, trans. by A. R. Price (Cambridge: Polity, 2014) p.162.
7. See Jacques Lacan, *The Seminar of Jacques Lacan, Book VII: The Ethics of Psychoanalysis, 1959–1960*, ed. by Jacques-Alain Miller, trans. by Dennis Porter (Abingdon: Routledge Classics, 2008) p.311. What is interesting here is that Lacan implies that it is not what is seen of the amorphous Thing itself, but its acting as an attractor, that best illustrates it: 'I am amazed at the murmur of pleasure that that name [Fellini] seems to have aroused among a significant number of you here today. I am ready to believe that this effect is only due to the moment of illusion produced by the fact that the things I say are calculated to emphasize a certain mirage, which is, in effect, the only one aimed at in the series of cinematographic images referred to. But it isn't reached anywhere except in a single moment.

That is to say at the moment when early in the morning among the pines on the edge of the beach, the jet-setters suddenly begin to move again after having remained motionless and almost disappearing from the vibration of the light; they begin to move toward some goal that pleased a great many of you, since you associated it with my famous Thing, which in this instance is some disgusting object that has been caught by a net in the sea. Thank goodness, that hadn't yet been seen at the moment I am referring to. Only the jet-setters start to walk, and they remain almost always as invisible, just like statues moving among trees painted by Uccello. It is a rare and unique moment.'

8. Thus, the subject's fantasy is:

$$\$ = S$$

but the structure of this fantasy is:

$$\$ \diamond a$$

The subject does not realise that seeking to be unbarred ('S') is to seek the object *a*, which stands in for the utterly unattainable.

9. Jacques-Alain Miller, 'Montré à premontré', *Analytica*, 37 (1984) pp.28–29, quoted and trans. by Slavoj Žižek in *Looking Awry: An Introduction to Jacques Lacan through Popular Culture* (Cambridge, MA: The MIT Press, 1991) pp.94–95.
10. Lacan, *Seminar X: Anxiety*, p.311.
11. Milner, 'Lacan and the Ideal of Science', in *Lacan and the Human Sciences*, p.29.
12. Ibid., p.31.
13. Lacan, *Seminar VII: The Ethics of Psychoanalysis*, p.15.
14. Malcolm Bowie in his *Freud, Proust and Lacan: Theory as Fiction* describes 'the knowledge-seeker's fictions and hypotheses [as] the mental constructions by which he experimentally produces the pleasures of knowledge as fine gradations of fore-pleasure' and summates how for the three protagonists of his work: 'against theory as fiction they set theory as anticipated certitude [the culmination of this fore-pleasure]. Freud in his archaeology, Proust's jealous narrator in his unwilled rememberings and Lacan in his mute encounters with the *réel* which forever lies beyond the reach of language, all have access to that 'joy resembling certainty' of which the narrator speaks. It is in the self-declaring play of desire between certainty and extinction, between bedrock and deadlock, that their new science begins.' See Malcolm Bowie, *Freud, Proust and Lacan: Theory as Fiction* (Cambridge: Cambridge University Press, 1987) p.166 and p.178.

CHAPTER 2

Monolith

Abstract Framing the eminently alienating and enigmatic monolith as an analogue of Freud's equally impenetrable primordial object, *das Ding*, or 'the Thing', this chapter investigates what cuts its appearances in *2001* brings into being. Giving a background to the making of the movie, and the writing of the novel, its sequels, and the screenplay (with Kubrick), by Arthur C. Clarke, the work of this chapter quickly turns to Lacan's disquisitions made on *das Ding* at the beginning of his seventh Seminar, *The Ethics of Psychoanalysis*. In so doing, it charts and traces the remarkable similitude of the monolith, as that which is incisive, and anxiety-provoking in that '*it does not deceive*', as Lacan put it in relation to *the Thing* in Seminar X.

Keywords Monolith • *das Ding/the Thing* • Stanley Kubrick • Arthur C. Clarke

It stands as something of a monolith of cinema. *2001: A Space Odyssey* was released in 1968, having grown out of a collaborative project between American filmmaker Stanley Kubrick and British science fiction writer Arthur C. Clarke, inaugurated in the Spring of 1964 when Kubrick contacted Clarke about the possibility of working on 'the proverbial 'really good' science-fiction movie' with him.[1] The latter left his home in the

D. Bristow, *2001: A Space Odyssey and Lacanian Psychoanalytic Theory*, The Palgrave Lacan Series,
https://doi.org/10.1007/978-3-319-69444-3_2

then Ceylon (Sri Lanka) to return to New York, both to editorially work on the nonfiction book *Man and Space* and to meet with Kubrick; it was in this city that they liaised and Kubrick suggested that they co-write a novel on which to base the screenplay.[2] Drawing on some elements of his 1948 short story 'The Sentinel', the novel took on a much-expanded shape from this point of departure under Clarke's composition, which continued into initial filming, although when it was finished Kubrick forbade publication of the book before release of the movie. The resultant film is an extraordinary artistic accomplishment, monolithic in its portrayals of space travel and exploration: its lengths of time, its isolations, its silence and vacuity, its returns of messages in inverted forms, and its surreal and defamiliarising aspects and affects.[3] The film set a precedent in cinema; indeed, Mark Kermode suggested somewhere that the original trailer for *Star Wars* (1977)—which George Lucas had claimed he had wanted to be like '*2001* meets James Bond, outer space and space ships flying in it'—closely mimicked *2001*'s cinematographic style, as it was initially thought that this would be the best method for attracting sci-fi audiences and would provide the way-in to the extraordinary success that the movie went on to achieve (which in the UK made it at that point in time the most popular film ever cinematically released, surpassing an attendance record set previously by *The Battle of the Somme* in 1916).[4] It was Kubrick's seventh feature film (of twelve proper), following on from and continuing certain themes of the masterful satire on the Cold War world *Dr. Strangelove or: How I Learned to Stop Worrying and Love the Bomb* (1964), and preceding the dark near-future dystopia of *A Clockwork Orange* (1971), adapted from Anthony Burgess' novel of 1962 (and officially unavailable in the UK from 1973 until Kubrick's death in 1999).[5]

Clarke wrote three sequels to *2001: A Space Odyssey*: *2010: Odyssey Two* (1982, which was made into a film by Peter Hyams in 1984), *2061: Odyssey Three* (1987), and *3001: The Final Odyssey* (1997); all of which, like the first, are far more explanatory than the film.[6] Although much can be made of them through dedicated literary studies, from this point onward any reference to the novel(s) will only be made secondarily to the film, as it is often precisely by putting the *explanatory* aside that interpretation can commence; that is, by piercing through the often all-too-convenient grand narrative—which in psychoanalysis may be likened to a cover story (such as that of the epidemic that is made necessitous to the crew of the Clavius Base in *2001*)—not by disbelieving it, but by interrogating its presuppositions, taking it back to its founding coordinates, or quilting points, and by prompting its rearticulation(s).[7] Thus, the divergences between the novel and the film will also not form a major consideration, in taking the movie as the primary object of inquiry, suffice here to mention the novel's ending, in which the Star Child—as the foetus depicted in the last scene of the film is called in the book—detonates gigantic explosive weapons (which is in fact what the first seen spacecraft in the film is) that are orbiting Earth. As Peter Krämer explains: 'despite his original intention to offer [this as] an optimistic ending, Clarke later acknowledged the validity of a pessimistic reading as well, thus affirming the fundamental ambiguity of the ending.'[8] Here, in effect acknowledging his *death as an author*—in the Barthesian sense that 'a text's unity lies not in its [psychobiographical] origins but in its destination', in its openness to interpretation—Clarke joined Kubrick, whose vision of *2001* he summarised as 'a 'nonverbal' experience'.[9] Instead, he 'tried to create a visual experience, one that bypasses verbalized pigeon-holing and directly penetrates the subconscious with an emotional and philosophic content'.[10] Rather psychoanalytically—and reinforcing the *interpretative link*—Kubrick also suggested that 'reactions to art are always different because they are always deeply personal[. T]he film becomes anything the viewer sees in it.'[11]

Over the course of what Krämer calls the 'transformation' of the making of the movie—that is, the collocation of alterations made during its gestation—this consideration seemed to become its guiding principle. Originally planned as a prologue to the film was a set of interviews with leading scientists of the day, discussing the possibility of extraterrestrial life, and other cutting-edge topics, and intended to add an air of scientific authority to the science fiction movie.[12] Although these interviews were recorded, the material fell away, as did the extensive voiceover narration that was written into the screenplay (as also occurred in the case of the final cuts of *Blade Runner*, made by Ridley Scott in 1982, and finalised by the director in 2007), the result of which was the non-narrative 'visual experience'—which becomes what the viewer sees in it—that Kubrick described.[13] This openness to interpretation would prevail to such an extent that towards the end of the shoot, contrary to the original plan of 'photographically depicting an extraterrestrial creature in a manner that would be as mind-boggling as the being itself', it was finally decided on that there would be no depicted extraterrestrials in *2001*, due to the fact that—as Kubrick puts it—'it soon became apparent that you cannot imagine the unimaginable'.[14] Thus, there is no alien lifeform imagined visually on behalf of the viewer; any such possibility is quite literally left only up to

Fig. 2.1 The appearance of the monolith on Earth four million years ago. Source: *2001: A Space Odyssey*

the imagination of the viewer, through its non-depiction (a cinematic convention famously and masterfully employed by Alfred Hitchcock, in particular in the shower stabbing scene in *Psycho* (1960), in which no flesh-piercing stabs are actually shown).

Thus, instead of 'little green men' we have the monolith—shown in Fig. 2.1 (in which it is apprehended by a group of apes whose reality will come to be fundamentally changed by it, as will discussed throughout this chapter)—concerning which (before getting to the *thing itself*) we will first deal with its mythologies: artistic, religious, sexual, metaphorical, and metonymical. In its blackness (the non-colour which Alain Badiou describes as 'passive negation[, as] it merely indicates the absence of its extreme opposite, light')—the monolith invokes a survey of the use of black in art and visual representation down through the ages, from being the primordial pigment used by the planet's first painters in the Palaeolithic era (via the materials of charcoal, or burnt bones)—through its being the basis for the ink of quills, pens, and printing presses in the revolutions of the dissemination of writing and printed illustration (the apogee of the latter of which perhaps being the deathly pages of black in Laurence Sterne's *Tristram Shandy* (1759–1767))—to the modernist and postmodernist experimentations, for example, of Kazimir Malevich's 'Black Square' (1915) or Mirosław Bałka's 'How It Is' (2009), which force confrontation with the starkness of blackness and darkness with as much enigmatic insistence as the monolith itself in *2001*.[15] When the rectangular ebon slab is depicted in the film it is thrice shown standing on one of the shorter congruent sides (as in Fig. 2.1, on Earth, although it is also seen floating in space between Jupiter and its moons horizontally—at least as far as camera angle is concerned, as space is without a universal axis; there is also little way of judging its size accurately in this scene due to the vastness of its perspective). Its positionality could thus be read (and indeed has been) as *phallic*, which—whilst this interpretation might present something of a goldmine for a certain type of lazy Lacanianism, or 'applied psychoanalysis'—is an analogy that perhaps should be met with the same incredulity of the apocryphal *dephallicising* riposte popular in the psychoanalytic imaginary: 'sometimes a cigar is just a cigar'.[16] Beyond appearing a few times standing erect, the eisegetic symbolism really stops there, especially if we take into account the variety of shapes, sizes, colours, and morphologies considered for the alien artefact throughout the creative transformation of the film, and the perfectly angularly symmetrical cuboid form settled on.[17]

More of a concern is the dominance of the *male* in *2001*, as a sex, as a gender—especially, that is, in a movie set in the future (although now past), in which gender roles could have been predicted to have been revolutionised (although—although it had modified—the gender bias was in no way overcome by the year 2001)—but this highlights much more issues with patriarchy than phallocentric privilege (symbolically and imagistically).[18] Yet, it is also the case that when Heywood Floyd (William Sylvester) boards Space Station V he meets with a group of four prominent Russian scientists, three of whom are women: his friend Elena (Margaret Tyzack)—whom he makes a point of shaking hands with before his Russian male counterpart, Dr Andrei Smyslov (Leonard Rossiter)—Dr Kallinan (Krystina Marr), and Dr Stretyineva (Irena Marr).[19] Whether or not this gender balance is a comment on the differences between predicted American and Soviet hierarchies in the future year can only be conjectural.[20] Again, material left on the cutting-room floor suggested alternatives too: the onboard computer aboard the *Discovery* was originally planned to be a female entity called Athena, for example; and, as Krämer recounts in relation to the ending to the film as it was planned out in an early version of the screenplay, Clarke commented:

> 'We never explained what happened to Bowman, but left it entirely to the imagination.' He suggested that instead of this open ending, the alien should give Bowman a spaceship, which is 'man's new tool—the equivalent of Moonwatcher's [the lead ape's] weapons', yet is much more feminine

('beautiful', 'soft', 'warm') than the phallic bones. According to Clarke, who wanted 'to push all sorts of subconscious and even Freudian buttons', the final voice-over should be 'Now he was master of the world.' Kubrick commented simply: 'I prefer present non-specific truth for film. Maybe this can work in a book but it wont [*sic*] on film.'[21]

More prominent than these gender disparities is of course the sparseness of humans in general in the film, and the subversion of the filmic consuetudes of the centrality of a single protagonist, or a love-interest-propelled plot or subplot; indeed, in *2001*, there is no sexual relationship. For a definitive reading of Kubrick's position on the (non-)sexual relationship and its vicissitudes—particularly in the modern cultural milieu—it is to Susan White that we will here turn, who in her essay 'Kubrick's Obscene Shadows' puts a key moment from *2001* alongside the (always already) constrained sexuality of Kubrick's last film, *Eyes Wide Shut*:

The final scene [of *Eyes Wide Shut*] contains Alice's imperative that she and Bill must "fuck" as soon as possible. But more important than this resexualization of marriage [...] is the fact that the couple and their daughter are shopping, looking at price tags, at the end of the film[.] Man, in *Eyes Wide Shut*, has evolved into *homo economicus*[.] Despite the participation of powerful men in orgies and the blood sacrifices of young women, obscenity is no longer in the shadows. It is out there for all to see; it is money in circulation, a postmodern emptying out of the culturally meaningful narratives of the past, without any resolution of the double binds posed by patriarchy and desire in Kubrick's earlier films. Kubrick's ambivalence toward cultural products ends here, with a deeply unsatisfying look at human platitudes and the shallowness of marital relations—or perhaps with a nod to the small satisfaction Daddy has in finally buying the bush baby for his little girl.[22]

This last sentence relates to the gift that Floyd's daughter asks of him for her birthday, via a videophone conversation between her on Earth and him on the space station. However, it is not confirmed in the final version of the film whether it is bought or not. (A scene depicting Floyd's purchase of it from an onboard Macy's was deleted, much to the company's chagrin.) What is implicitly teased out by White is the inextricability of the reproduction of capitalism and of the species (which itself has become '*homo economicus*'); it is economic circulation that has the totemic hold over the modern *epistēmē*.

Thus, we might read the linking of phallicism to the monolith as in fact something of a reverse-transference onto it from the bone—which precedes the cut from the Pleistocene to the Space Age—with all the common sexual metaphoricity that goes with it. Indeed, the monolith itself disappears from Earth and its cut is forgotten by the early apelike ancestors of humankind, although the result of the cut is not, and becomes symbolised in the bone, which is thus given a sense of originariness. Comparative religion presents instances of bones involved in creation and regeneration; for example, the creation of woman (Eve) from the rib-bone of man (Adam) in the Judeo-Christian heritage (described in the Bible at Gen. 2.22), or that of the luz, a spinal bone, which in Judaic and Islamic traditions was believed to be indestructible and from which it was believed that the body could be rebuilt at the Resurrection.[23] Perhaps the epitomic statement of an osseous origin is the contradiction highlighted by Hegel: '*Spirit is a bone.*' This famous phrase is a truncation of a passage of what could be called Hegel's *phrenomenology* in the *Phenomenology of Spirit*—that is, his phenomenological discussion and critique of the pseudoscience phrenology, which is predicated on the idea that mental capacity is derivable from measurements taken of the skull—which it will be worth quoting from here at length:

> To say that Spirit [merely] *is*, means nothing else than that it is a Thing. When *being* as such, or thinghood, is predicated of Spirit, the true expression of this is that Spirit is, therefore, the same kind of being that a *bone is*. It must therefore be regarded as extremely important that the true expression has been found for the bare statement about Spirit—that it *is*. When in other respects it is said of Spirit that it *is*, that it has *being*, is a *Thing*, a single, separate *reality*, this is not *intended* to mean that it is something we can see or take in our hands or touch, and so on, but that is what is *said*; and what *really* is said is expressed by saying that the *being of Spirit is a bone*.[24]

In effect, in his analysis of this passage, J. N. Findlay relates the movement in it back to a cut, in which recognition by the observation of the contingency of inner and outer has taken place (in Lacanian terms, this may be seen as the operation—even the inauguration—of the Imaginary order) and thus 'the reality of Spirit [i]s thereby made into a thing, and inert being given the significance of Spirit.'[25] Thus, the parallelism of old (the theory of mind and body running separately but simultaneously, in parallel, as propounded by the likes of Leibniz) gives way here to dialectical materialism proper, and to the recognition of an inextricability that both *unifies and separates at once*: e.g., bone and Spirit are separate in their individual *thingness*, but unified in *being*.[26] (Thus, these considerations can in effect be transposed to the interplay between the three Lacanian orders—the Real, the Symbolic, and the Imaginary—one of the ways of doing which is by tracing the extensions of Hegelianism in Karl Marx's work, specifically here in the discussion of the economic credit system in the third volume of *Capital*, in which Marx states that 'with the development of the credit system, capitalist production constantly strives to overcome [it]s metallic barrier [that is, that of metal money, the gold standard, etc.], which is both a material and an imaginary barrier to wealth and its movement, while time and again breaking its head on it.'[27] To be both material (real) and imaginary at once can be construed as the founding act of the (latent) Lacanian unconscious, which comes into (manifest/speaking/alterative) being through symbolisation (much like the credit system; after other exchange and monetary systems).[28] It comes into its hypostasis (what Lacan calls '*parlêtre*'; literally 'speakingbeing') by its entry into the symbolic realm, which is outside of itself, and therefore excessive to it, the interplay of all of which—if left unchecked (for example, by psychoanalysis, if not by other, personal or institutional, psychological means)—can lead to unique modes of functioning, management, or coping; to symptomatic occurrences; and to haywire excesses (much like the credit system and its resultant crashes and crises). Thus, these symptomatologies of the Symbolic form *the social field*—through a *discursive link*—constitutive of the unconscious as it ex-sists (as it marks the distinction outside/inside). To return to the (metallic) 'barrier' in Marx, he also states that 'a drain of gold, therefore, shows strikingly by its effects that production is not really subjected to social control, as social production, and that the social form of wealth exists alongside wealth itself as a *thing*.'[29] What *really* is wealth is supported, *shown*, by the symbolic (and here real and symbolic are imaginarily registered, or connected). Thus, the apperception of the (real) *Thing* can only come through the apprehension of the (symbolic) *thing*.)

As a proof of the phrenological fallacy (which misunderstands the element of Spirit in its material determinism, and which resonates with Moon-Watcher's—the lead 'man-ape', unnamed in the film, and played by Daniel Richter—*sublated* use of the bone), Hegel suggests that the logical retort to the phrenologist's suggestion that "you (your inner being) are this kind of person because your skull-bone is constituted in such and such a way" would of necessity 'have to go the length of beating in the skull of anyone making such a judgement, in order to demonstrate in a manner just as palpable as his wisdom, that for a man, a bone is nothing *in itself*, much less *his* true reality.'[30] Indeed, in *Less Than Nothing* Žižek returns us to the fact that the formula '*Spirit is a bone*' is proposed by Hegel 'as the absolute contradiction, not as a little bit of the real constitutive of subjectivity.'[31] Thus, for this *little bit of the real* it is the cut that must be gone back to, behind which lies the Thing, which we will make an attempt to gravitate towards in the remainder of this chapter (despite the roadblocks inherent to the excavation and isolation of this primally 'profound' and 'ultimate' enigmatic object).

* * *

'Well, there it is.'[32] These are Dr Ralph Halvorsen's words that serve as the only descriptor for the monolith as it stands in the Tycho crater on the Moon in the *2001* screenplay (Fig. 2.2 depicts this environment in the film, in which the monolith stands uncovered in the middle, and in the foreground the scientists are seen approaching it, as Earth shines behind in the sky). In Clarke's novel, it receives this introduction at that same point in the story:

> A vertical slab of jet-black material, about ten feet high and five feet wide; it reminded Floyd, somewhat ominously of a giant tombstone. Perfectly sharp-edged and symmetrical, it was so black it seemed to have swallowed up the light falling upon it; there was no surface detail at all. It was impossible to tell whether it was made of stone, or metal, or plastic—or some material altogether unknown to man. [...]

Fig. 2.2 The monolith—known as TMA-1—standing upright in the Tycho crater excavation site. Source: *2001: A Space Odyssey*

> At some signal, floodlights around the lip of the crater were turned on, and the bright earthlight was obliterated by a far more brilliant glare. In the lunar vacuum the beams were, of course, completely invisible; they formed overlapping ellipses of blinding white, centred on the monolith. And where they touched it, its ebon surface seemed to swallow them.
>
> Pandora's Box, thought Floyd, with a sudden sense of foreboding—waiting to be opened by inquisitive Man. And what will he find inside?[33]

Of course, in the film, it (or its perfect likeness) has appeared before—at the dawn of man, some four million years prior—but it does not appear *as such* at that point in either the screenplay or the book. In the screenplay, the apes encounter a cube fifteen feet across and 'made of some completely transparent material', which overtly becomes a teaching machine, projecting images within itself, as if the apes were watching a theatrical production taking place in a glasshouse.[34] In the novel, in place of the first monolith as it appears in the film, Moon-Watcher comes face to face with 'a rectangular slab, three times his height but narrow enough to span with his arms, and it was made of some completely transparent material' (just like the abovementioned cube).[35] It does not teach as explicitly as the screenplay's cube, but fills with manifestations of lights and instantiates a more abstract tutelary experience to which the apes unrealisingly submit and from which the same change in the course of their history as in all the three versions results. In the film, the monolith is rectangular, pitch-black, and diegetically mute (although the 'Kyrie' section of György Ligeti's *Requiem* gives its presentation such an urgent sense of awe and ominousness).

At about twelve minutes into the film, the monolithic thing makes its first appearance, throwing the apes into a state of frenzied curiosity; they make tentative approaches, and touch it unsurely. As the music rises the sun also rises above the monolith, in line with the fainter outline of the moon. At the music's zenith, the film then cuts to the silence of the ancient African landscape. After a few exterior shots, the apes are seen ambling around by a skeletal collection of bones; Moon-Watcher scratches about and—as it dawns on him to take a bone in hand—the shot of dawn over the monolith is cut back to briefly. Moon-Watcher proceeds to break up the animal skeleton with his newly fashioned club; as he smashes the skull, there are two cutaways to a slayed tapir falling to the ground. A cut has thus been instantiated.

We then see the apes eating meat, the food they had not realised had been wandering around in close proximity to them all the while as they nearly starved from their vegetative diet. After that, a fight ensues between the apes and a rival unenlightened gang, and Moon-Watcher clubs the opposition's leader to death, striking fear into the hearts of his subordinates, before throwing his bone-weapon into the air. The scene cuts to the craft floating off of Earth four million years in the future.

Lacan's first thesis in his 1948 paper 'Aggressiveness in Psychoanalysis' is that '*aggressiveness manifests itself in an experience that is subjective in its very constitution.*'[36] In the man-apes of Clarke and Kubrick's prehistoric Earth an experience of aggressiveness is manifested by the monolith, constitutive of their subjectivity. What might seem strange, from a Lacanian point of view, is that any semblance of the coming-into-being of the very field and function of speech and language is elided in the famous filmic four-million-year cut.[37] Despite the fundamentality of speech and language (to subjectivity), the experience of aggressiveness here thus in effect predates them. Indeed, the '*specific prematurity of birth*' that Lacan mentions in 'The Mirror Stage as Formative of the *I* Function' can also be related to this very birth of humankind; that is, before entry into the symbolic order and the communication it gives rise to, for as Lacan puts it in Seminar X[38]:

> At root the subject receives his own message from the Other. The first emergence, the one that is set down in the table, is simply an unconscious, since it is unformulable, *What am I?*, to which corresponds, before it is actually formulated, a *Thou art*. In other words, the subject first receives his own message in an inverted form.
>
> I've been saying this for a long time. Today, I'm adding, if you care to lend an ear, that he first receives it in an initially broken-off form. First of all he hears a *Thou art* without any attributive. As broken-off as this message is, however, and therefore insufficient, it is never unformed, because language exists in the real, it is afoot, in circulation, and many things [...] have already been settled in this language.[39]

What is here shown to be primordial is the sending of the prelinguistic (as *structured like a language* as it might be) message '*what am I?*', and formative of the *I* function is its eventual return in its inverted form—'*Thou art*'—by the Other, here allegorised by the monolith (as that primordial object, the Thing. Indeed, as Derek Hook states of *das Ding*: 'while Lacan clearly *does* want to understand the relations of the subject to something primordial[,] he nevertheless wishes to avoid the fixity of a singular, "nuclear" object and its variants. He prefers instead to provide an operational definition of the field of *das Ding* [*the Thing*] which establishes the framework for such relations'[40]). The emergence of this initial exchange (or framework of relations) is thus as what Lacan calls in Seminars XVI and XVII a *discourse without speech*.[41] What the film (over the written material) further demonstrates is the broken-offness of this first message, given as it is by the stonefaced enigma of the monolith, which indeed presents an 'enigmatic signifier', in Jean Laplanche's phrasing.[42] It doesn't inaugurate speech (which could be construed as its 'insufficiency'), but rather makes a subjective cut *on the way to speech*, as—as Lacan puts it—language, existing in the real, 'is afoot, in circulation', structuring, laying the foundations, or cutting the pathways of discursive relations (transmitting the communicative conduits of the field of the Other). It would be metaphysical hermeneutics to speculate on whether the monolith already knew of the specificity of this circulation, or whether it was the bringer or harbinger of this element of the real, etc.[43] Thus—in line with an explanation that Lacan delivers in 'Aggressiveness in Psychoanalysis'—what the monolith *qua* Thing seems to introduce by its sudden appearance at the dawn of man is a 'formal fixation, which introduces a certain difference of level, a certain discordance between man as organism and his *Umwelt* [the world as it is experienced by him], [and which] is the very condition that indefinitely extends his world and his power, by giving his objects their instrumental polyvalence, as well as their potential as weaponry.'[44]

The concept of 'the Thing' is derived from Freud's usage of the term '*das Ding*' in his *Project for a Scientific Psychology*, completed in 1895, but first published in 1950 in a collection edited by Marie Bonaparte, Anna Freud, and Ernst Kris called *Aus den Anfängen der Psychoanalyse*, the translation of which by James Strachey as *The Origins of Psycho-analysis* in 1954 was the *Project*'s first appearance in English. Elucidating Lacan's above statement from 'Aggressiveness in Psychoanalysis' perfectly, Freud delineates in the *Project* how in the human organism 'perceptual cathexes coincide in part with information from one's own body. As a consequence, the perceptual complexes are divided into a constant, non-understood, part—the *thing*—and a changing, understandable, one—the attribute of movement of the thing.'[45] The misrecognition, or non-understanding, by what Freud would come to term 'the perception-consciousness system' (or 'system Pcpt.-Cs.') of the Thing becomes the motor drive of its cut (its movement, or the subject's moving towards (refinding) it).

Lacan's erudite and in-depth disquisition on *das Ding* begins Seminar VII, *The Ethics of Psychoanalysis*. He explains therein that 'the *Ding* is the element that is initially isolated by the subject in his experience of the *Nebenmensch* [neighbour] as being by its very nature alien, *Fremde*.'[46] In *2001*, the monolith has appeared to, and neighbours, the primal horde of the apes; it is what they initially isolate, and thereafter—through repetition of its cut (enacted over time and evolution with an increasing metonymical stretch of its signification across the chain of signifiers)—becomes the object that they perpetually try to refind; that is, even after any recollection of the monolith has completely amnesiacally disappeared.[47] The Brazilian psychoanalyst Noga Wine thus summarises the subjective process that has taken place and its axiomatic points in her key study, 'The Object from Freud to Lacan':

> Lacan suggests that as *das Ding* itself cannot be represented, it will always be represented as "another thing" or "something else", to use Freud's words. The object is already different from the Thing, and each object which represents *das Ding* will represent it by means of *difference itself*[.] No empirical object could fill in the place of *das Ding*, as there is no possibility to fill the void it has left. It is the movement from one object to another which indicates *das Ding* as an exterior motor which motivates the change from one to the other.[48]

This void-Thing, and its driving force, is encapsulated in the featurelessness of *2001*'s monolith, the absence of comprehension of which in the screenplay is demonstrated by Halvorsen's words, lacking any description beyond indication, which themselves go unsaid (by Robert Beatty) in the film, silence speaking volumes in their place. In its subjective formulation, confrontation with the void of the Thing motivates the drive of endless circulation around that void—around *the Thing* itself—in which the ineluctable election of stand-in after stand-in for the original Thing thus occurs, none of the objects of which will attain to its strange dignity (in Lacan's winking choice of the word '*dignité*' the pathway of course opens for its anagrammatic punning rearrangement into *Dingité*), although attainment to it would be beyond the realm of the reality principle (not to mention the pleasure principle), as perhaps only Bowman (fictionally structured as he is)—in the whole history of the subject—knows, although this knowledge itself is a knowledge *beyond knowledge itself*.[49] Thus, the 'instrumental polyvalence' that is given to objects in the experience of subjectivity that Lacan comments on in 'Aggressiveness in Psychoanalysis' is perhaps just this; that is, their (attempted) refinding. (As alluded to above in relation to Marx, the apperception of the real comes at the expense of its severance from (impossible) direct contact; that is, through the mediation of the symbolic (which cannot—constitutively, structurally—offer up the (noumenal) real itself, but only a (phenomenal) representation of it), creative of the circulatory drive of desire, around the real; as *the Thing*, and in its 'little bits'. Indeed, as Lacan puts it—after Hegel—in 'The Function and Field of Speech and Language in Psychoanalysis': 'the symbol first manifests itself as the killing of the thing, and this death results in the endless perpetuation of the subject's desire.'[50])

The recurrences of the monolith and its avatars at various points in *2001* will be dealt with within the remainder of this work. However, of the monolith as the Thing—described so appropriately by Clarke in *2010* as 'the very symbol of mystery and unattainability. For it has never been touched; it can only be worshipped from afar'—after a short detour, we will leave the last words to Lacan.[51] As Dylan Evans puts it: 'after the seminar of 1959–60 [*The Ethics of Psychoanalysis*], the term *das Ding* disappears almost entirely from Lacan's work'—much as it does from Freud's, and much as the Thing itself does from Moon-Watcher's and subsequent humanity's *Umwelt*—although it 'provide[s] essential features' that will be continually returned to (in the *objet petit a*, for example).[52] In the words of Jacques Derrida: 'the Thing [*Chose*] haunts[;] it causes, [and] it inhabits without residing'.[53] Such can only be the true functioning of *das Ding*, for as Lacan states definitively in Seminar VII:

> The whole progress of the subject is [...] oriented around the *Ding* as *Fremde*, strange and even hostile on occasion, or in any case the first outside[.] It is, of course, clear that what is supposed to be found cannot be found again. It is in its nature that the object as such is lost. It will never be found again. Something is there while one waits for something better, or worse, but which one wants.
>
> The world of our experience, the Freudian world, assumes that it is this object, *das Ding*, as the absolute Other of the subject, that one is supposed to find again. One doesn't find it, but only its pleasurable associations. [...]
>
> [Thus,] the world of perception is represented by Freud as dependent on that fundamental hallucination without which there would be no attention available.[54]

That is, the symbolised imaginary hallucination of the real of the Thing.

Notes

1. For a chronology, see Carolyn Geduld, 'The Production: A Calendar', in *The Stanley Kubrick Archives*, ed. by Alison Castle (Cologne: Taschen, 2016) pp.400–417, and also p.394 of this work, which includes a passage from the initial letter that Kubrick sent to Clarke, from which the quotation is taken.
2. For Clarke's account of this correspondence and the beginnings of the making of *2001*, see Clarke, 'Foreword to the Special Edition', in *2001: A Space Odyssey*, pp.xix–xxviii.
3. Initially for Joseph Gelmis, the *longue durée* of space travel and exploration in *2001* was too excruciatingly slow for his first-time-round sensibilities, and he reflectingly offered the summation—a convolution of Marshall McLuhan's famous catchphrase—that he believed that in the movie 'the tedium was the message'. This was stated in an article on his return to *2001*, after he had panned the film upon its initial release. For the first review, and his subsequent revisions of his opinion of the film, and Kubrick, see Joseph Gelmis, 'From *Newsday* (April 4, 1968): *'Space Odyssey' fails most gloriously*'; 'From *Newsday* (April 20, 1968): Another look at 'Space Odyssey''; and 'From *Newsday* (April 5, 1969): *Understanding the message of* 2001', in *The Making of Kubrick's* 2001, ed. by Jerome Agel (New York: Signet, 1970) pp.263–265; pp.265–268; and pp.268–269, respectively. The above quotation is on p.267. See also Joseph Gelmis, *The Film Director as Superstar* (Harmondsworth: Pelican, 1974), in which he extols the virtues of Kubrick's directorial prowess.
4. George Lucas, quoted in Larry Sturhahn, 'The Making of *American Graffiti*' [1974], in *George Lucas: Interviews*, ed. by Sally Kline (Jackson: University Press of Mississippi, 1999) p.32; and Peter Walsh, 'Screening Notes to *The Battle of the Somme*', pamphlet distributed at a 2016 screening at The Little Theatre, Bath, collaboratively produced by the Imperial War Museums and Bath & North East Somerset Council. *Star Wars* can be seen as the movie which—leaving aside earlier flying saucer B-movies and their ilk—moved the mainstream genre of the science fiction film both from the symphonic to the operatic and from the universal/subjective to the factional/alterative. Representing the birth of the blockbuster, it was thus the founder of the long-lasting hegemony of these motifs in mainstream cinema (which nonetheless were a continuation from other genres, such as that of the Western, for example. Indeed, Žižek highlights the Western's 'sublation' in the space opera. See Slavoj Žižek, 'Class Struggle or Postmodernism? Yes, Please!', in Judith Butler, Ernesto Laclau, and Slavoj Žižek, *Contingency, Hegemony, Universality: Contemporary Dialogues on the Left* (London: Verso, 2000) pp.99–100. Parallels between

2001 and *Star Wars* and the psychoanalytic methodologies of Freud and Lacan and the psychiatric theory of C. G. Jung could of course also be made here.

5. Kubrick's twelve feature films, after the disowned *Fear and Desire* (1953), are thus: *Killer's Kiss* (1955), *The Killing* (1956), *Paths of Glory* (1957), *Spartacus* (1960), *Lolita* (1962), *Dr. Strangelove or: How I Learned to Stop Worrying and Love the Bomb* (1964), *2001: A Space Odyssey* (1968), *A Clockwork Orange* (1971), *Barry Lyndon* (1975), *The Shining* (1980), *Full Metal Jacket* (1987), and *Eyes Wide Shut* (1999). He had originated and worked on the concept for *A.I. Artificial Intelligence* (2001) with Steven Spielberg, which the latter directed, dedicating the movie to the former. *A Clockwork Orange* was withdrawn at Kubrick's request, after it was blamed for copycat violence, which resulted in direct threats to and protests against the director and his family. However, Kubrick did not recant on his conviction that the accusations had got the dialectic between art and life quite the wrong way round.
6. See Clarke, *2001: A Space Odyssey*; Arthur C. Clarke, *2010: Odyssey Two* (London: Voyager, 1997); Arthur C. Clarke, *2061: Odyssey Three* (London: Voyager, 1997); and Arthur C. Clarke, *3001: The Final Odyssey* (London: Voyager, 1997). In *2010*, the space race ramps up between the newly united USA and Russia, and the Chinese. Returning to Jupiter and the *Discovery*, Heywood Floyd joins HAL's creator Dr Chandra (who is named Mr Langley in the film of *2001*) and other crew members, and witnesses the ingestion of Jupiter by self-replicating monoliths to create a new miniature sun called Lucifer, Jupiter's moons becoming worlds for humans to explore (apart from Europa, which is to be avoided at all costs). *2061* begins with a centurion Floyd travelling to Halley's Comet aboard the celebrity vessel the *Universe*; the plot collides with the alien warning about Europa at the end of *2010*, and unravels over the millennium and into the cumulative finale of *3001*, in which—in either a paean to a discouraged Kubrickian idea, or through forgetfulness or intentional contravention—Clarke animates an idea he had fought hard to abandon in 1965, as this diary entry indicates: 'March 8. Fighting hard to stop Stan from bringing Dr. Poole back from the dead. I'm afraid his obsession with immortality has overcome his artistic instincts.' See Clarke, *The Lost Worlds of 2001*, p.36. Interestingly, for the sequels, Clarke generally followed the plot of the film of *2001*, as opposed to the book's, (in which, for example, the action takes place in Saturn space as opposed to Jupiter space; however, reference made in *2010* to the Stargate sequence and the fate of Dave Bowman is more faithful to the written version of *2001*). He also later wrote a trilogy of *Time Odyssey* books with Stephen Baxter: *Time's Eye* (2003); *Sunstorm* (2005); and *Firstborn* (2007).

7. Quizzed himself on the point of explanatoriness, Kubrick said in an interview with Joseph Gelmis in the Spring of 1968: 'Once you're dealing on a nonverbal level, ambiguity is unavoidable. But it's the ambiguity of all art, of a fine piece of music or a painting. You don't need written instructions by the composer or painter accompanying such works to explain them. "Explaining" them contributes nothing but a superficial cultural value which has no usefulness except for critics and teachers who have to earn a living. Reactions to art are always different because they are always deeply personal.' See Stanley Kubrick, quoted in Piers Bizony, *The Making of Stanley Kubrick's* 2001: A Space Odyssey (Cologne: Taschen, 2015) p.424.
8. Peter Krämer, *2001: A Space Odyssey* (London: Palgrave Macmillan/British Film Institute, 2010) p.16. The optimistic reading entails the Star Child destroying humanity's means of destruction (of itself), which also produces energy for the Star Child's consumption, according to the plot of *2010*; the pessimistic, that the destruction of the weapons at the novel's end is harmful to both the planet and its life.
9. See Roland Barthes, 'The Death of the Author' [1967], in *Image Music Text*, ed. and trans. by Stephen Heath (London: Fontana, 1997) p.148, and Stanley Kubrick, quoted in 'Stanley Kubrick: *Playboy* Interview', in *The Making of* 2001: A Space Odyssey, ed. by Stephanie Schwam (New York: Modern Library, 2000) p.272.
10. Ibid. As Jean Laplanche and Jean-Bertrand Pontalis explain, 'subconscious' is a 'term used in psychology as a designation for what is scarcely conscious or else for what is below the threshold of immediate consciousness or even inaccessible to it. Used by Freud in his earliest writings as a synonym for 'unconscious', it was very quickly discarded because of the confusion it tends to foster.' See Jean Laplanche and Jean-Bertrand Pontalis, *The Language of Psychoanalysis*, trans. by Donald Nicholson-Smith (London: H. Karnac, 1998) p.430. Kubrick appears to be using the term in the sense of the unconscious, as was and is a common traduction, especially amongst American references to the psychoanalytic concept.
11. Stanley Kubrick, quoted in Alison Castle, 'Preface', in *The Stanley Kubrick Archives*, p.7. The word 'link' is used above in the precise Lacanian sense that entails a discursive relation. James Joyce said something much the same of his modernist masterpiece *Ulysses* (1922): 'though people may read more into *Ulysses* than I ever intended, who's to say that they are wrong: do any of us know what we are creating?' See Arthur Power, *Conversations with James Joyce*, ed. by Clive Hart (Chicago: University of Chicago Press, 1974) p.103. Of course, *Ulysses* (as its title suggests) derives its plot from Homer's *Odyssey* (Ulysses being the Latin name for the Greek character Odysseus). *2001* itself transposes many elements of the terrestrial odyssey into space (as *its* title suggests); however, mapping Homeric or

Joycean parallels onto the film is beyond the remit of this study, and thus we might leave this fragmentary commentary of the elusive (but not so much in reverse) Nahpets Vorodoch to suffice: '*Odyssey* = the journey of Odysseus (Ulysses) who in the Joyce version at least meets his 'true' son (Stephen) whose name is Dedalus (and Icarus, flying beyond man's limits, etc.). Also, meeting one's self = transport of self back to childhood (end of film). Mabbott Street = atmosphere of Jupiter (with all those silly blinking oil-slicks)?? (Well, the chaos before realization works in both anyway.) Wild drinking at end of Mabbott Street—and glasses break! One of Joyce's fun words is metempsychosis. (Molly pronounces it met-him-pike-hosses? [*sic*]) = 'rebirth of souls,' certainly descriptive of end of film.' See Nahpets Vorodoch, quoted in *The Making of Kubrick's* 2001, p.302.

12. For the partially edited transcripts of these, see *The Making of Kubrick's* 2001, pp.27–57. For more information on the process of their collation, see Anthony Frewin, 'Stanley Kubrick and the Search for Extraterrestrial Intelligence', in *The Stanley Kubrick Archives*, pp.418–424.
13. On *Blade Runner*, and its different versions, see Matthew Flisfeder, *Postmodern Theory and Blade Runner* (London: Bloomsbury, 2017).
14. Stanley Kubrick, quoted in *The Stanley Kubrick Archives*, p.424.
15. Alain Badiou, *Black: The Brilliance of a Non-Color*, trans. by Susan Spitzer (Cambridge: Polity, 2017) p.34. In a phrase precursory of the inscrutable but transformative effect of the monolith, Malevich says of his painting, in a Suprematist manifesto of 1927: 'The square seemed incomprehensible to the critics and the public … and this, of course, was to be expected.' See Kazimir Malevich, 'Suprematism', in *MANIFESTO! A Century of isms*, ed. by Mary Ann Caws (Nebraska: University of Nebraska Press, 2001) p.405. For more on Bałka's piece—an enterable pitch-black oblong space—see my piece 'Mirosław Bałka's "How It Is"', in EDA Collective, *Why are Animals Funny? Everyday Analysis, Volume 1* (Alresford: Zero Books, 2014) pp.105–108, and Zygmunt Bauman and others, *Miroslaw Balka, How It Is*, ed. by Helen Sainsbury (London: Tate Publishing, 2009).
16. This comment cannot be directly traced to Freud, but seems to have been first attributed to him by Allen Wheelis in his 1950 paper 'The Place of Action in Personality Change', *Psychiatry*, 13th ser., 2 (1950) 135–148 (p.139, note 9), in which he dates it to 1920. In Seminar XXIII Lacan is posed the question: '*Is your twisted cigar a symptom of your real?*' and replies (no doubt with Freud's cigar—and its status as *just that*—in mind): 'It certainly is. My twisted cigar bears the strictest possible relation to the question that I posed about the equally twisted straight line of the same name.' See Jacques Lacan, *The Seminar of Jacques Lacan, Book XXIII: The Sinthome*, ed. by Jacques-Alain Miller, trans. by A. R. Price (Cambridge: Polity, 2016) p.120. He discusses the bending of a straight line earlier in

the Q & A, in relation to a metaphor incorrectly attributed to V. I. Lenin concerning economists having 'bent the stick' too far one way, and the necessity of having to bend it back (pp.117–118). The cigars that Lacan smoked were *culebras*, which came in packs of three and were bent around each other, and which—if their curvature is followed through, to being tied up end to end, forming rings—comes close (but no cigar) to a Borromean link, the topological formation that is privileged in the twenty-third Seminar, and in the late Lacan more generally. This tying-up process in fact makes a link known as 6^3_3 in the Rolfsen notation (whereas the Borromean link is 6^3_2). In Seminar IV, Lacan does highlight the structure of 'pure and simple erection. A stone elevated to a vertical position is one of its examples, the notion of the human body in so far as it is erect is another.' See Jacques Lacan, *The Seminar of Jacques Lacan, Book IV: The Object Relation, 1956–1957*, ed. by Jacques-Alain Miller, trans. by L. V. A. Roche (Unofficial translation, n.d.) p.48. Jung goes so far as to make of cigar-shaped UFOs a sort of (phallic) archetype in Carl Gustav Jung, *Flying Saucers: A Modern Myth of Things Seen in the Sky* [1958], trans. by R. F. C. Hull [uncredited] (London: Routledge Classics, 2002) p.103; see also p.14, n.6. An example of a Lacanian reading of *2001* that relies on phallicism (reading the monolith's symbolism thusly, and the bone as being acoustically close to 'boner', which strikes a rather strange note in the prelinguistic sonority of its setting) can be found in Wim Matthys, David Schrans, and Filip Geeradyn's '*2001: A Space Odyssey* and the Anticipatory Function of the Symbolic Order in the Staging of the Beating Fantasy', *International Journal of Computing Anticipatory Systems*, 29 (2014) 69–75 (p.73), in which the authors even find the director's name in the monolith, as 'Cube-brick'; why stop, however, when this dubious signifier's reach—why not?—might even be extended to 'Cube-prick'?

17. In the novel, the first monolith is transparent and crystalline, and also becomes 'suffused with a pale milky luminescence'. See Clarke, *2001: A Space Odyssey*, p.13. In the screenplay, it is at first a cube. In this, there is perhaps a metonymic link with the Ka'ba, the holy building in the great mosque Al-Haram to which the pilgrimage to Mecca is made by Muslims, and towards which prayer is turned, and of which Clarke makes mention in *2010* (p.177). This is a rectangular building covered in a black shroud, the cornerstone of which, known as the Black Stone, is thought to be of meteoric origin, and is encased in a silver oval embrasure, offsetting the structure's angularity. On the speculated origin of the stone, see Elsebeth Thomsen, 'New Light on the Origin of the Holy Black Stone of the Ka'ba', *Meteoritics*, 15th ser., 1 (1980) 87–91. Indeed, later, Clarke said of it: 'though I certainly did not have it in mind at the time, the fact that the Black Stone sacred to Moslems is reputed to be a meteorite is more than

quaint coincidence.' See Arthur C. Clarke, quoted in *The Making of Kubrick's* 2001, p.290.

18. For a critical interaction with the sexual politics of *2001*—which frames the film in a favourable light compared to the gendered inscriptions of other prior sci-fi—see Barry Keith Grant, 'Of Men and Monoliths: Science Fiction, Gender, and *2001: A Space Odyssey*', in *Stanley Kubrick's 2001: A Space Odyssey: New Essays*, ed. by Robert Kolker (New York: Oxford University Press, 2006) pp.69–86. Beyond this also is of course the issue of its all-white cast. This point, however, was raised in a prescient review of the film on its release—and one of the only positive ones—by Penelope Gilliatt, which saw such (futurally continuous) tropes as intentionally placed (however rooted its language is in a racialist past): 'the [human] race has survived thirty-three years more without extinction [that is, since 1968], though not with any growth of spirit. There are no negroes in this vision of America's space program; conversation with Russian scientists is brittle with manly terror, and the Chinese can still be dealt with only by pretending they're not there.' See Penelope Gilliatt, 'After Man', *The New Yorker*, April 13, 1968, quoted in Bizony, *The Making of Stanley Kubrick's* 2001: A Space Odyssey, p.418.
19. The Marr sisters' roles were uncredited.
20. Notwithstanding a possible continuation of the 'pinko' politics of *Dr. Strangelove*, perhaps into a division between a communistic gender utopianism and an imperial patriarchy, wishing to extend itself across the universe. (Dan Blocker purportedly described the script of *Dr. Strangelove* as 'too pinko' for him to consider playing the role of Major T. J. "King" Kong in the movie, and relayed this through his agent to Terry Southern. See Terry Southern, 'Strangelove Outtake: Notes from the War Room', *Grand Street*, 49 (1994) 64–80 (p.71). In *2010*, the Russian crew of the *Leonov* is captained by a woman, and is made up of two women to three men.
21. Krämer, *2001: A Space Odyssey*, p.47. For information on Athena, see p.30. According to Geduld, this version of the computer was decided on by Clarke for the novel on August 6 1964, and presumably subsequently revised in the following March (see 'The Production: A Calendar', in *The Stanley Kubrick Archives*, pp.404–405) and became the basis for the computer in the first draft of the screenplay when the working title for the film was *Journey Beyond the Stars* (completed 24 December 1964).
22. Susan White, 'Kubrick's Obscene Shadows', in *Stanley Kubrick's 2001: A Space Odyssey: New Essays*, p.143.
23. For a detailed survey of references to this bone (which is the coccyx), see Robert Shapiro, 'The Mystical Bone of Resurrection', *Radiology*, 163rd ser., 3 (1987) 718. In Seminar XVIII Lacan elevates writing to the status

of bone, around which the flesh of language forms: 'writing, from its origins, up to its last protean techniques, is only something that is articulated as bone (*os*) of which language is the flesh.' See Jacques Lacan, *The Seminar of Jacques Lacan, Book XVIII: On a Discourse that Might Not Be a Semblance, 1971*, trans. by Cormac Gallagher (London: Karnac, unofficial translation, n.d.) session IX, p.6 [trans. modified]. Cormac Gallagher's translations tend to follow the format of giving the number of the Seminar session, followed by the page number, beginning from the start of each session. For more on this comment from Lacan, see Daniel C. Bristow, 'Seminars XVII and XVIII with XXIII: On a University Discourse that Might Not be Joycean', *The Letter: Irish Journal for Lacanian Psychoanalysis*, 61 (2016) 45–75 (p.48).

24. Hegel, *Phenomenology of Spirit*, p.208. The parenthesis is Miller's.
25. J. N. Findlay, 'Analysis of the Text', in Hegel, *Phenomenology of Spirit*, p.540. Further to this analysis, in *God in Pain* Žižek proposes two possible initial readings of the truncation '*Spirit is a bone*': 'the vulgar materialist "reductionist" reading—the shape of our skull effectively and directly determines the features of a man's mind—and the speculative reading—the spirit is strong enough to assert its identity with the utmost inert stuff and to "sublate" it, i.e., even the utmost inert stuff cannot escape the Spirit's power of mediation'. See Slavoj Žižek, 'A Glance into the Archives of Islam', in Slavoj Žižek and Boris Gunjević, *God in Pain: Inversions of Apocalypse* (New York: Seven Stories Press, 2012) p.114. Thus, in the movement of sublation (*Aufhebung*), Spirit cancels-yet-preserves the bone, rendering it *dead* (the bone as symbolic of death) whilst still *being* (the bone, like Spirit, nonetheless *is*).
26. Perhaps a key to disambiguation in English is a linguistic one, between the nounal and verbal states; that is to say, Spirit and bone are not two types of being that are the same, but are of the same type of being, in that each *is*. In other words, it is not that there is Being A (e.g., Spirit) and Being B (e.g., bone), which are the same (they indeed have their own *thingnesses*), but that Being A and Being B are the same, in *being* (as in the verb *to be*). In the original German, the necessity of the definite article possibly makes definition clearer in this respect.
27. Karl Marx, *Capital: A Critique of Political Economy, Volume III* [1894], trans. by David Fernbach, 3 vols (London: Penguin Classics, 1991) III, p.708.
28. By 'alterative' is here meant that which is both (actively *and* passively) altering and *other* at once.
29. Marx, *Capital: Volume III*, p.708.
30. Hegel, *Phenomenology of Spirit*, p.205.

31. Slavoj Žižek, *Less Than Nothing: Hegel and the Shadow of Dialectical Materialism* (London: Verso, 2012) p.457.
32. Stanley Kubrick and Arthur C. Clarke, *2001: A Space Odyssey, Screenplay* (Borehamwood: MGM-British Studios, 1965, photocopy) p.b62, dated '12/14/65'. The pagination in the screenplay relates to the sections of the film, and each section is dated with when it was composed, which indicates where certain insertions were made.
33. Clarke, *2001: A Space Odyssey*, p.69 and p.78.
34. Kubrick and Clarke, *2001: A Space Odyssey, Screenplay*, p.a14, 10/13/65. When filming had begun, Kubrick's idea was to make the monolith that the apes behold a translucent rectangular slab. After the prop had been arduously created out of a massive Perspex block, it was rejected by Kubrick. This block became the material out of which the sculptor Arthur Fleischmann made the artwork *Crystal Crown* for Queen Elizabeth II's Silver Jubilee, which was unveiled in 1977 and is displayed to this day at St. Katherine's Docks, London, with a plaque indicating its Kubrickian origins. See Bizony, *The Making of Stanley Kubrick's* 2001: A Space Odyssey, p.182.
35. Clarke, *2001: A Space Odyssey*, p.10.
36. Jacques Lacan, 'Aggressiveness in Psychoanalysis' [1948], in *Écrits: The First Complete Edition in English* [1966], trans. by Bruce Fink, in collaboration with Héloïse Fink and Russell Grigg (New York: W. W. Norton, 2006) p.83.
37. When we do arrive in the *speaking* socius of 1999 (the year of Floyd's trip to the moon), we are met with a banality of unemotive, technical language and a dry type of precise but disengaged speech (all of which is much commented-upon in *2001* criticism), which chimes with what Lacan 'define[s] as that which is satisfied by blah-blah'—empty though authoritative phraseology—in Seminar XX. See Lacan, *Seminar XX: Encore*, p.56.
38. Jacques Lacan, 'The Mirror Stage as Formative of the *I* Function as Revealed in Psychoanalytic Experience' [1949], in *Écrits*, p.78.
39. Lacan, *Seminar X: Anxiety*, p.272.
40. Derek Hook, 'The Powers of Emptiness', *Theory & Psychology*, 20th ser., 6 (2010) 855–870 (p.861).
41. On this, see Jacques Lacan, *The Seminar of Jacques Lacan, Book XVI: From an Other to the other, 1968–1969*, trans. by Cormac Gallagher (London: Karnac, unofficial translation, n.d.); Jacques Lacan, *The Seminar of Jacques Lacan, Book XVII: The Other Side of Psychoanalysis* [1969–1970], ed. by Jacques-Alain Miller, trans. by Russell Grigg (New York: W. W. Norton, 2007); as well as Daniel Bristow, *Joyce and Lacan: Reading, Writing, and Psychoanalysis* (Abingdon: Routledge, 2016) pp.125–168.

42. On this point—which will be returned to in the coming chapter—I owe thanks to discussion with Andrew Bryant.
43. Certainly in the novel, and increasingly so in its sequels, the monolith and its avatars do have a certain foreknowledge that relates to an ancient exploratory expedition to Earth.
44. Lacan, 'Aggressiveness in Psychoanalysis', in *Écrits*, pp.90–91.
45. Sigmund Freud, '*Project for a Scientific Psychology*' [1895], trans. by James Strachey, in *The Standard Edition of the Complete Psychological Works of Sigmund Freud, Volume I (1886–1899): Pre-Psychoanalytic Publications* and *Unpublished Drafts*, ed. by James Strachey, with Anna Freud, assisted by Alix Strachey and Alan Tyson, 24 vols (London: Vintage, 2001) I, p.383. The *Standard Edition* of Freud's works will hereafter be referred to as *SE*, followed by volume then page number.
46. Jacques Lacan, *Seminar VII: The Ethics of Psychoanalysis*, p.62. Of the word '*Nebenmensch*', Lacan says: 'the formula is striking to the extent that it expresses powerfully the idea of beside yet alike, separation and identity' (Ibid.).
47. In the novel, Clarke describes this amnesia: 'Moon-Watcher paused uncertainly at a familiar spot. Something, he knew was missing, but what it was he could not remember. He wasted no mental effort on the problem'. See Clarke, *2001: A Space Odyssey*, p.26.
48. Noga Wine, 'The Object from Freud to Lacan', in *Introductory Lectures on Lacan*, ed. by Astrid Gessert (London: Karnac Books, 2014) p.49.
49. In his formula for sublimation in Seminar VII, Lacan states: 'it raises an object—and I don't mind the suggestion of a play on words in the term I use—to the dignity of the Thing.' See Lacan, *Seminar VII: The Ethics of Psychoanalysis*, p.138. The dialectic of the beyond will be returned to in the last chapter.
50. Jacques Lacan, 'The Function and Field of Speech and Language in Psychoanalysis' [1953], in *Écrits*, p.262.
51. Clarke, *2010: Odyssey Two*, p.293.
52. Dylan Evans, 'Thing', in *An Introductory Dictionary of Lacanian Psychoanalysis* (London: Routledge, 1996) p.208. The Thing does reemerge in Lacan's subsequent Seminars, but nothing like to the extent with which it is dealt in VII; as Lacan alludes to, in Freud the concept of *das Ding* is referred to in the article on 'Denegation', as well as in the *Project*.
53. Jacques Derrida, *Specters of Marx: The State of the Debt, The Work of Mourning, and The New International* [1993], trans. by Peggy Kamuf (Abingdon: Routledge Classics, 2006) p.21.
54. Lacan, *Seminar VII: The Ethics of Psychoanalysis*, pp.62–63.

CHAPTER 3

Cut

Abstract This extensive chapter on the cut launches an exploration into Lacan's career-long reliance on topological modelling, which culminated in a surfeit of the practice in his late Seminars, which were almost conducted without words and whilst 'Lacan was no longer anything but thought', as he has been described by François Cheng. The chapter files the cuts of *2001* into the categories of 'major' (each monolithic cut: inaugurating hominid, then computer, sentience, and those that beckon the *beyond*) and 'minor', and relays the cuts found in the formations of the topologies of the cross-cap and Borromean linkages against the 'Freudian structures' of neurosis, perversion, and psychosis, and their dialects. A heuristic topological map is laid out, which it is hoped offers bearings for future explorations.

Keywords Cut • Topology • Cross-cap • Borromean links • Alexandre Koyré • Freudian structures

The subject, the Cartesian subject, is what is presupposed by the unconscious[.]
The Other is the dimension required by the fact that speech affirms itself as truth.
The unconscious is, between the two of them, their cut in action.
—Jacques Lacan, 'Position of the Unconscious' [1960, rev. 1964], in Écrits, *p.712*

D. Bristow, *2001: A Space Odyssey and Lacanian Psychoanalytic Theory*, The Palgrave Lacan Series,
https://doi.org/10.1007/978-3-319-69444-3_3

To begin with a coda to the preceding chapter: strictly speaking, a monolith cannot be (a) cut, but it can instantiate a cut (by definition, a cut implies (at least) two—certainly a multiple—whereas what is mono is always one). At this point we might take after Lacan, in 'The Instance of the Letter in the Unconscious, or Reason Since Freud', in which he regales his readership with a tale about two children trying to work out from two signs that they see at a station whether their train has arrived at the destination 'Ladies' or 'Gentleman', and states that 'the rails in [this] story materialize the bar in the [reversed] Saussurian algorithm' of signifier over signified: $\frac{S}{s}$ (a bar which later in the *Écrits* he labels a 'cut').[1] That is; visually, we can perhaps suggest here that the monolith is able to 'materialise' the (epistemological) cutting manoeuvre itself, if pictured side-on, at a diagonal ('/'); i.e., it can materialise as the bar that divides the subject (creating: '$\$$', the divided subject). This materialised splitting bar is thus brought about by how the object cuts into the subject, and how it thereby founds the fundamental fantasy, as outlined earlier in the 'Overture'; the formula of which is: '$\$ \diamond a$'.

In an extended footnote added in July 1966 to the *écrit* that summarised his teaching in the *Psychoses* Seminar (III) of 1955–1956—'On a Question Prior to Any Possible Treatment of Psychosis'—Lacan states that he situated the object *a*, in its position in a certain schema that he depicted in the body of the text, 'so as to shed light on what it contributes regarding the field of reality (a field that bars it[; the *a*, that is]).'[2] It is situated in the 'R schema', certain elements of which we will now isolate and investigate herein Fig. 3.1.

In the footnote, Lacan draws attention to 'the only valid cut in this schema (the cut $\overrightarrow{mi}$, $\overrightarrow{MI}$)', which delimits the shaded area, '*R*' (by following

Fig. 3.1 The 'R' Schema, featured in the *écrit*: 'On a Question Prior to Any Possible Treatment of Psychosis'. Source: Jacques Lacan, *Écrits*, p.462

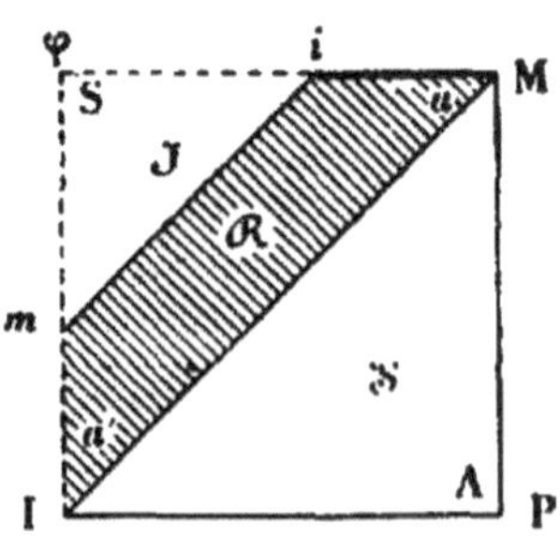

the lines between the positions *m* and *i*, and M and I, and joining them up, forming a trapezium, in two dimensions)—and highlights its structure as a Möbius strip (in three dimensions, directionally denoted by the arrows above '*mi*' and 'MI')—containing the object *a*, and its avatar, *a'*.[3] Thus, we see that *das Ding* has now boiled down to its compulsive stand-in for the subject, the object *a*, and its repetitions (*a'*, *a"*, *a"'*, etc.). According to Lacan: 'this says it all, since this field [$\overrightarrow{mi}$, $\overrightarrow{MI}$] will henceforth be the mere placeholder of the fantasy whose entire structure is provided by this cut', by which he means:

> Only the cut reveals the structure of the entire surface, because it is able to detach from it the following two heterogeneous elements (noted in my algorithm ($\$ \lozenge a$) of fantasy): $\$$—the barred S of the strip to be expected here where it in fact turns up, that is, covering the field ***R*** of reality—and *a*, which corresponds to the fields ***I*** and ***S.***
>
> It is thus as representation's representative in fantasy—that is, as the originally repressed—that $\$$, the barred S of desire, props up the field of reality here; and this field is sustained only by the extraction of object *a*, which nevertheless gives it its frame.[4]

We are here returned to the frame as Miller depicts it in his 'Montré à premontré' that was drawn upon in the 'Overture'. The extraction of object *a* gives the field of reality—in its disjunction from the Real—its very sustenance (that is, as the fantasy which covers over the Real, and blocks or bars (access to) the impossible *real* of the object *a*). As Lacan delineates, it is only the cut that is able to reveal the structure itself, in that it 'detaches' its (two) 'heterogeneous elements' (in this case). That is; whilst linking them, it separates them out too, just like a cut does two scenes in a film. In this respect, the cut operates along the lines I have elucidated elsewhere of the *unificatory/separatory principle*, which denotes that which links *and* separates *at once* that which is cut (positions 1 and 2: the Real and reality, or, in the formula of the fantasy, $\$$ and *a*), which is brought about *by this cut itself* (position A; the cut, or the lozenge in the fantasy formula: $\lozenge$). In the instance of the fantasy it is thus demonstrable as[5]:

◊

(cut)

/ \

$\$$ *a*

Thus, the object that originarily founds the fantasy is *das Ding*, which is endlessly attempted to be refound in a succession of objects (*a*, *a'*, *a"*, etc.), the attempts at the promotion of which to its status are endlessly made by the subject. Whilst such attempt at promotion is how Lacan formulises sublimation, it also underwrites the functioning of the drive. For example, as he explicates in Seminar IV—in counterpoint to the theories of the Objects Relations school of psychoanalysis (the lead exponents of which were Melanie Klein and D. W. Winnicott, and which often took the mother to be the primordial object)—in 'the first Freudian dialectic of the theory of sexuality', it is the 'lack of accord', between what is sought and what is (re)found, in relation to the primordial object, that keeps the drive desirously recirculating (through libidinal stratification, or by alighting on a single stand-in that arrests the fascination and disavows its object-cause (or semblance thereof)—as in the fetish—or through a myriad of other sexual meanderings):

> There is always an essential division, fundamentally conflictual, in the refound object, and by the very fact of its refinding there is always a lack of accord in the object that is refound with regard to the object that is sought. That is the notion with which the first Freudian dialectic of the theory of sexuality is introduced.
>
> This fundamental experience supposes that there is, in the course of the latency period, a conservation of the object in memory, unbeknownst to the subject, that is to say, a signifying transmission. This object then enters into discord and plays a disturbing role in every later object relation of the subject. It is in this framework, at certain selected junctures and at certain times of evolution, that the strictly imaginary functions are revealed. Everything arising from the pregenital relation is taken into this parenthesis. Into a dialectic which is first, essentially, in our vocabulary, a dialectic of the symbolic and the real, the imaginary layer is then introduced.[6]

Thus, in terms of the cut *as the unconscious*—and the effects of cuts *as unconscious*; that is, of the unconscious as *the enactment of a cut*, to paraphrase the Lacan of the *écrit* 'Position of the Unconscious', quoted at the beginning of this chapter—what is visualised by Kubrick in *2001: A Space Odyssey* is the very 'conservation of the object' and its 'signifying transmission' in the memory. Retained in the conservation by transmission is thus not the object itself, *as signified*, but *as signifying*; that is, not its *meaning*, but its *address*. To return to the primordial framework of relations inaugurated by *das Ding*, as discussed in the previous chapter; in Laplanche's theory of the 'enigmatic signifier' (as derived from Lacan), he states that there is 'the possibility that the signifier may be *designified*, or lose what it signifies, without thereby losing its power to signify *to*'—and thus, in extension, it is not only that the signifier may be(come) *designified*, but that it could be primordially *designifiant* (as in the case of *das Ding*).[7] Those that have come into encounter with the monolith have thus been signified *to*, although enigmatically (i.e., what the signification is *of* proves utterly elusive); so much so is the signification beyond comprehension that only its traces can imaginarily repeat—in a sort of Möbian manner (which will be returned to momentarily)—in the unconscious and its manifest parapraxes, as endlessly in search of completion in a solid meaning—which will inevitably only elude the grasp of the understanding—as they are.

Thus, in *2001*, once the epistemic cut has been instantiated, and its effects become inscribed, its traces symbolically repeat throughout the film, in its symbolic unconscious. For example, the shape and hue of the monolith reappear in shadowy doorframes, or in the HAL9000 computer's main control panel; the thrown bone-weapon transitions to an orbiting weaponised spacecraft, and then to a banally bureaucratic pen floating in zero-gravity aboard a commercial space flight (with all the Arendtian weaponisation that such a symbol might suggest). If this is thus the dialectic of the real and the symbolic that we are articulating, it is the imaginary that has (subjectively) connected things up, its functions being revealed in the memorisation of the first symbolic impressions spawned from the real, that have gone on to spur their attempted replication in the emblemata of representation. (If, previously, we suggested that a linking of the Real and the Imaginary produces the *unconscious* (in something like its *tabula rasa* state), which hypostasises through symbolisation (connection with the Symbolic, of the world), it is here the imaginary that brings the *subject* into its constitution, through interaction with its *Umwelt*—in which real and symbolic pre-exist it in a linked relation—which connects the subject to, and disentangles it from, the *Umwelt*'s resonances, unificatory/separatorily.)

To thus articulate the functions of the 'imaginary layer'—as Lacan calls it—in a final demonstration of this topology (before moving into the 'Borromean' realm), we will turn to the strange structure of the cross-cap, a shape made up of a Möbius strip—the section which we could roughly say forms the 'cross'—and a domed disk, or half-sphere—the section forming the 'cap'—deformed to connect to it in such a way as to extend the traversability of the strip. All of this, on—or rather *as*—the *real projective plane*.

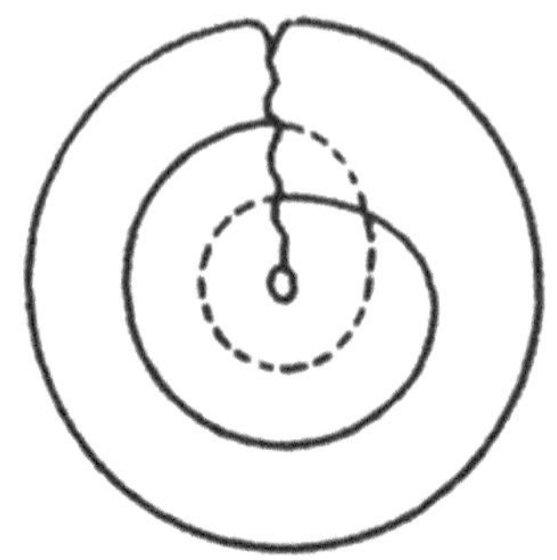

Fig. 3.2 The cross-cap. (Top half: Möbius strip; bottom half: domed disk.) Source: Jacqueline Rose, Sexuality in the Field of Vision, p.189

Which is to say: due to the fact that the Möbius strip of the cross-cap intersects itself, it is non-orientable in three dimensions; thus, it would require a fourth to realise its dimensionally abstract properties. If properly completed, or conceptualised, therefore, it is a single-sided shape (and thus strict dissection into its elements is dependent on its dimensional manifold.)

In *Sexuality in the Field of Vision*, Jacqueline Rose explains its Lacanian applicability, and gives a depiction of the shape (Fig. 3.2). (Its construction begins by taking an annulus and making a radial cut in it; the ends of this split are then self-intersected to make a non-orientable Möbius strip, which is attached to the half-sphere forming its bottom). As she puts it:

> The model used is the cross-cap or projective plane. By means of a cut the model can be split into two separate parts, one of which will seize its central point and the other of which will appear as a Moebius strip. The latter now represents specular identification, the former the subject in its relation to desire. The model is difficult, but two basic factors should be retained:
>
> *(a)* the cut which constitutes the subject in its dependent relation to the object of desire also allows the subject to *detach* itself from the specular illusion;
> *(b)* the cut which detaches this fragment also determines the topological properties of the fragment which remains; hence the specular illusion as apparently separate, but always the *effect* of the basic structure of desire: 'the essential factor is that the *coupure* [cutting] which detaches the object is that which simultaneously determines the topological properties of the fragment which remains'.[8]

To transpose the structure to the Lacanian orders of the Real, Symbolic, and Imaginary, in a brief digression—as is being done to a certain extent in Rose—we will utilise the below diagrammatic demarcations (in Fig. 3.3). We can locate the 'cross'—the Möbius strip section—on the plane of the

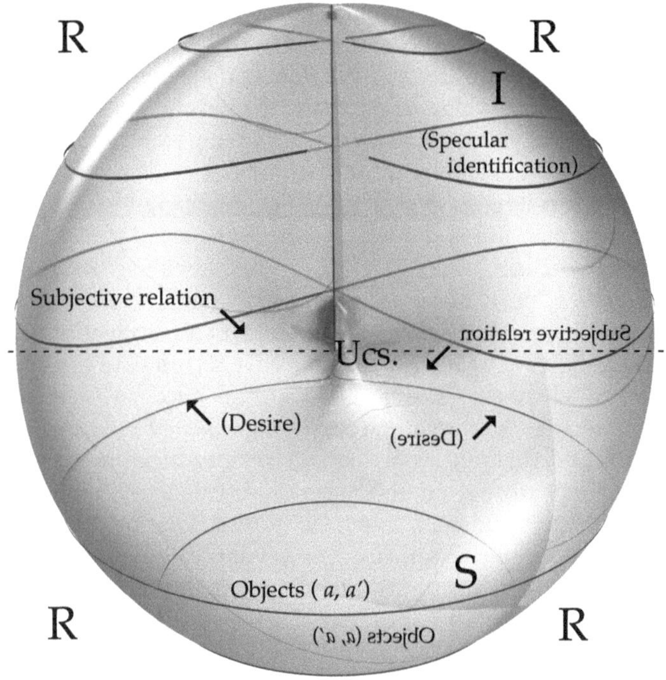

Fig. 3.3 Cross-cap depicting the three orders' structuration thereof; positions of the unconscious, its relations, and its objects; and directions of desire. Source: the author

Imaginary ('I'; that of Rose's 'Specular identification') and the 'cap'—deformed domed disk section—on that of the Symbolic ('S'; where the 'Subjective relation' intersects with 'Desire', the *dialectic* of which will be described below). The structure of the cross-cap itself thus resides in the Real ('R'), conveniently, here, an *unrealisable*—that is, non-transposable to (our) *reality*—projective plane. We could say, it is circumscribed by the Real (hence the four 'R's 'around' the cross-cap in the diagram), and we need not say that it circumscribes it 'inside and out', as its inside is its outside.

According to Rose; unifying and separating the two elements of 'cross' and 'cap' (in 3-space, that is) is thus the unconscious, *as cut*. (Further, then, in this model, the unconscious acts as unifier/separator of the three orders themselves, in their specificity; that is: of the Möbial structure, for the Imaginary; of the (semi-)sphericity of the domed disk, for the Symbolic, on which can be found symbolic objects—only subjectively accessible through the extension of the Imaginary Möbius strip's traversability—and of which objects we can never grasp behind or beyond them the *Thing* itself (i.e., from the outside, *without thus being on the inside at the same time*); and of the Real's all-encompassing inaccessibility (its strange structure as constitutively *non-all*: its structural circumscription and utter subjective removal). Interestingly, if the unconscious cut here brings the three orders together, it does so in something like the operation of the *sinthome* in the late Lacanian topology: thus, as the *sinthome* of the existential (or dimensional) situation.)

The *dialectic of desire* in the diagram is shown in how desire feeds to the specular subject (in the Imaginary)—*effecting* it and its identifications (*à la* Rose's description)—and in how the subjective relation traverses desire and its objects, or object-causes: objects *a* (*a*, *a'*, *a"*, etc.). We could call the first-listed direction (as they are properly *simultaneous*) the 'desirous relation of subject-formation', and mark it on the diagram with upward-pointing diagonal arrows, and its dialectical reverse the 'subjective relation of desire-formation', and mark it with downward-pointing diagonal arrows. Thus, coming out into the Möbial extension of the domed disk, the subjective relation of desire traverses the environs of the objects it symbolises ('refinds'), but can only thereafter re-enter the Möbial cycle (remobilising desire, *identificatorily*; i.e., subjective desire then identifies with certain objects). All of this process, of course, rather than taking hold of *das Ding*, which itself could be imagined as the separated-off deformed domed disk (if the marked line of the cut is followed: '- - - -'), in its *thingness*; that is, as *real* and eminently *symbolisable*, but subjectively so only through the *imaginary*, which, in subjectively connecting it, fundamentally disconnects it from the subject, hence the 'lack of accord' in the refound object, and the eternal returns of desirousness. Thus, that (always subjectively already) eviscerated gap between the Möbius strip ('cross') and the half-sphere ('cap') becomes the cut of the unconscious: its enacting of the dialectic of desire, and the repetition compulsions of trauma; its trying to get back to the *Thing*, the loss of which it enigmatically encircles.

* * *

The major cuts of *2001* are brought about by the monolith, and propel not only the film's plot, but—within its universe—the very advancements of sentience itself: thus, they are when the monolith first appears on Earth, as apprehended by the apes; its excavation on the Moon, and its disorienting transmission (after which we are party to a new era in AI sentience); its appearance before Bowman's journey beyond the stars, and the infinite, through the Stargate; and its quiescent yet transformative presence at the end of Bowman's bed in the surreal hotel setting at the film's end. Other prominent cuts in the film might thus be construed as minor in comparison to what establishes as major those just mentioned, some of these being the jump-cut from bone to spacecraft; the cut to the 'Jupiter Mission: 18 Months Later' title-card, after the sequence on the Moon; the film's intermission, a turning point in relation to the HAL computer (voiced by Douglas Rain) and its breakdown; and the cutting off of Bowman's sole crewmate—waking one, that is, as three other scientists are initially hibernating onboard—Frank Poole (Gary Lockwood), from his oxygen supply by an unmanned spacepod controlled by HAL; and indeed each instance of death in the film (of which there is a surprisingly high number—between animal, hominid, human, and computer—somewhat ironically, given its Universal (U) rating by the British Board of Film Classification).[9]

In the following, we will work heuristically towards new definitions and topological modellings of cuts. What is a major cut may be defined—in Lacan's words in 'The Subversion of the Subject and the Dialectic of Desire in the Freudian Unconscious'—as one 'made by the signifying chain[, which] is the only cut that verifies the structure of the subject as a discontinuity in the real.'[10] The signifying chain is (an element of) the symbolic, which gets seized upon by the imaginary. To return to the 'R' schema; as the cut reveals the structure of the fantasy, covering *over* the real, it also verifies the subjective position *in* the real; that is, as a discontinuity within it. Thus, that the real is (imaginarily) punctuated by the symbolic allows Lacan—in 'The Situation of Psychoanalysis and the Training of Psychoanalysts in 1956'—to extrapolate from archaeological excavations the assertion that:

> No prehistory allows us to efface the cut brought about by the heteronomy of the symbolic. On the contrary, everything it gives us merely deepens the cut: tools whose serial form directs attention more toward the ritual of their fabrication than toward the uses to which they were put; piles that show nothing other than the symbol anticipating the symbolic's entry into the world; and graves which, beyond any explanation that we can dream up for them, are edifices unknown to nature.[11]

In ruins thus remain runes, the sedimented symbols of the Symbolic. It is due to the Symbolic's 'heteronomy' that symbols can thus (retroactively) anticipate their symbolisation: the Symbolic—its heteronomy—(ostensibly) cuts into the Real, and is connected up by the Imaginary.

A taxonomy of attributes might here be plottable against the three Lacanian orders, if we now phrase things in a different topological register to that used above (culminating in the cross-cap): if *heteronomy* defines the Symbolic, then *autonomy* might properly be attributable to the Imaginary (as the realm of the *autos*; that is, of the ego, of the *I*, and of (primary) narcissism: of the *mirage* (*mir*ror-im*age*) of (the) self). For the Real we might break, or agglutinate—indeed, both, *at once*—this binary with the interjection of *homonomy*, taken here to mean the unnamed/unnameable: the indistinct, undifferentiated, or uncut. Following the above Lacanian (chrono)logic, the process of the three orders' interlinking—and thus also the process of subjectivation itself—can be presented in three stages. (That is, for the purpose of clear diagrammatisation only, we are here dividing a spontaneous phenomenon into sequentiality; a non-temporal sequentiality, that is.) This three-stage process can further be presented in two ways, which I will—however provisionally—here label 'general' and 'processual':

General (Fig. 3.4):

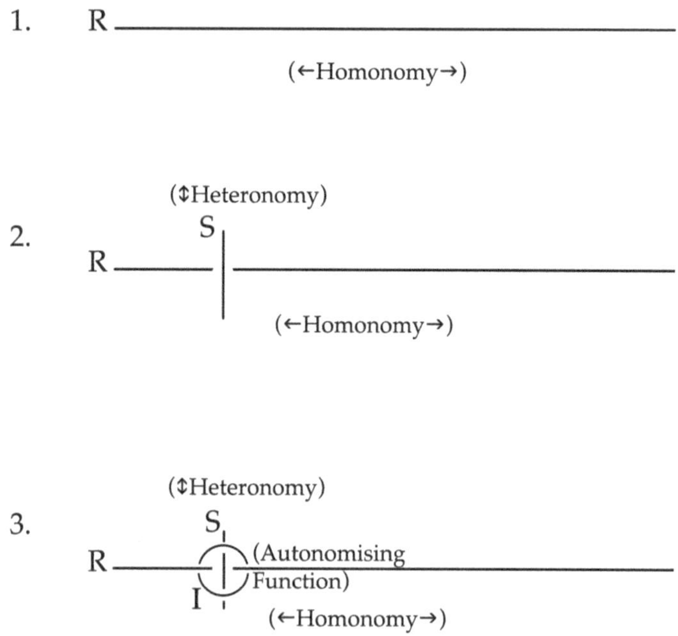

Fig. 3.4 'General' homonomy-heteronomy-autonomy schema. Source: the author

Processual (Fig. 3.5):

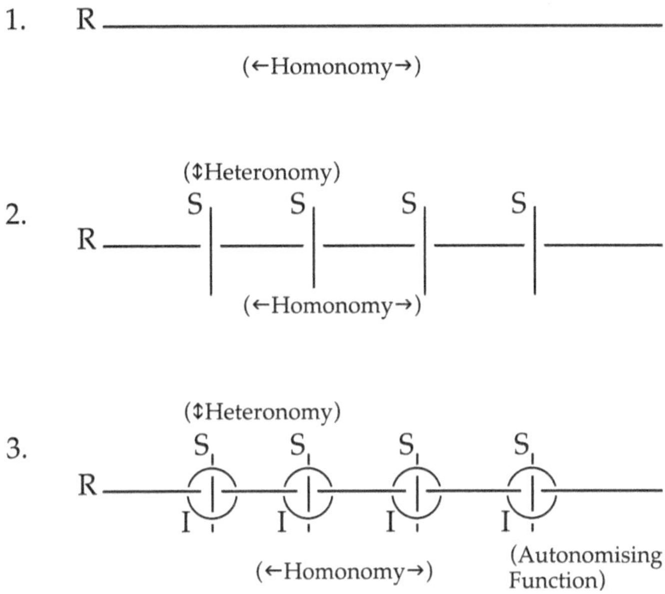

Fig. 3.5 'Processual' homonomy-heteronomy-autonomy schema. Source: the author.

To begin with the 'general' presentation:

1. If the Real ('R') is taken as an infinite line, its *homonomy* can be seen as continuous and consistent. We thus depict this here on the horizontal axis.
2. Thus, taken also as an infinite line, the Symbolic ('S') introduces its *heteronomy*, and *appears* to make a cut in the Real. However, this is not actually the case, due to the fact that:
3. The 'autonomising function' of the Imaginary ('I')—its bringing together (unifying) and differentiating (separating) the Real and Symbolic, through its *autonomy*—confirms this tripartite structure as the form known as 'the circle and the cross' (Fig. 3.6), which is on the way to depicting a Borromean link (and of course displays a certain affinity with the structuration of the cross-cap).

Fig. 3.6 The circle and the cross. Source: the author

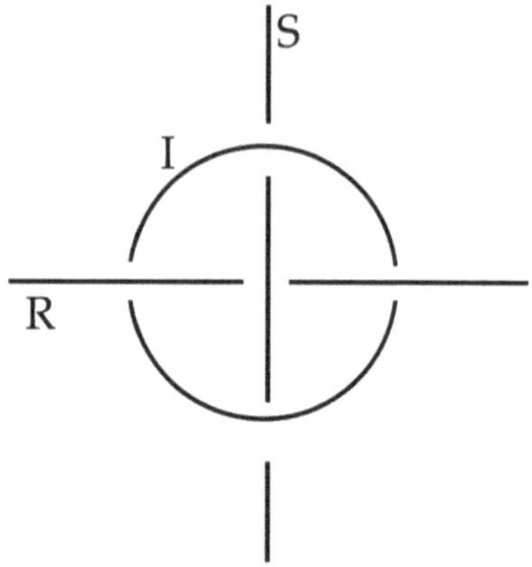

This structure is, strictly speaking, only *on the way* to depicting a Borromean link, for—whilst it is true that if any one of its components were to be cut (the ring, or either of the infinite lines), the structure would fully unravel (that is, no two of the other components would remain connected)—whereas the Borromean link has six crossing points, this permutation has only five: four on the circle ('I')—at its intersections with the two axes ('R' and 'S')—and that of the intersection of the two axes of the cross.[12] The vertical and horizontal lines thus need to join end to end respectively, over the existing ring, to themselves become rings (which partake of the property of infinity, in their circularity), which will create the sixth crossing point and therefore the Borromean link.

Thus, whilst the heteronomous process involved in stage 2 initially *appears* to make a cut in the Real, it in fact rather *displaces* the Real, which *brings about* the true cut—that is, the *hole*, at the centre of the Borromean link—only through the third stage; that of the structure's topologisation, enabled by its unificatory/separatory tripartition. This is seeable when given in the structuration of the Borromean link (Fig. 3.7).

The 'general' presentation thus gives us the formation of the interlinking of the Real, the Symbolic, and the Imaginary, as three continuous and consistent planes, and as the three constituents of subjective psychality; their separateness *from* yet unification *with* each other being the unconscious. If it is this unificatory/separatoriness that is their *cut*, it is the *hole* (formed from the displacements of the orders by their force upon each other) that this Möbian cut encircles (as the place perhaps of the *Thing*; of Freud's *navel*; of the object *a*, etc. Indeed, it is here that Lacan places the *a* in his

Fig. 3.7 Borromean link. Source: the author

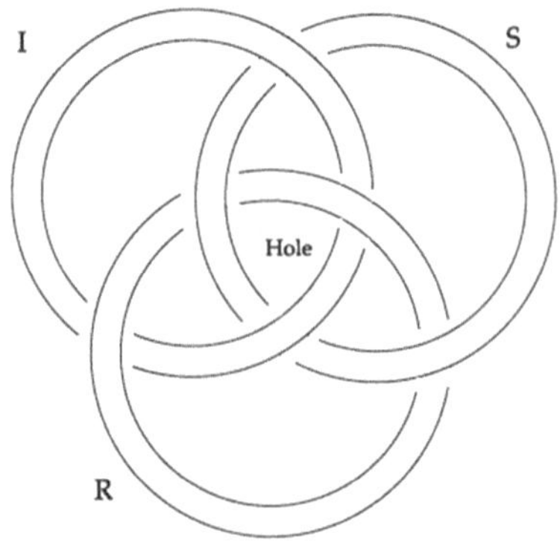

late work). As the 'general' presentation, this can be seen to represent the basic situational, or existential, topology of the three orders; constitutive, that is, of the *Umwelt*, and of the unconscious. The (properly processual) signifying chain itself would in this depiction necessitously be inscribed within the ring of the order of the Symbolic (i.e., if the ring is imagined as a tyre, the chain could then be imagined as an inner tread, for example).

Thus, the 'processual' presentation separates the signifying chain out into its Symbolic moments or irruptions; into signifiers, that is. Ergo:

1. The continuous and consistent line of the *homonomy* of the Real ('R') remains the same as in the 'general' presentation at this stage.
2. The Symbolic process of 'signifierisation' (signifiers—that is, 'S's—proliferating) is *heteronomously* introduced.
3. Each is connected up, and thus demonstrated to be separate—that is, *autonomised*—by the Imaginary ('I') function.

If the infinite lines of 'R' and 'S' were again connected end to end in the same manner as above in the 'general' presentation, interesting linkages emerge; that is, the ring of the Real becomes the sustainer of a series of Borromean linkages to it, as in Fig. 3.8.

Thus, if the ring of the Real were cut, each of the Imaginary and Symbolic links would come away entirely; however, if one of the Imaginary or Symbolic links were to be cut, it would only effect the dehiscence of the particularised link of which it is a constituent part (that is, the dehiscence of the particular *autonomised signifier*). In terms of the 'process' of the signifying chain, things can thus be apprehended as so: a little bit of

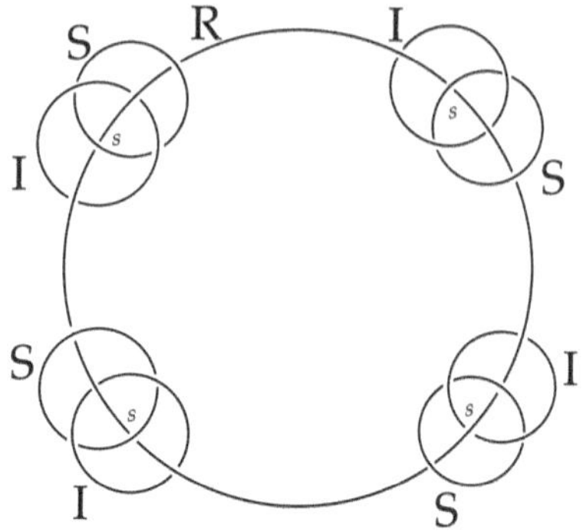

Fig. 3.8 'R'-ring with Borromean linkages. Source: the author

the infinity of real is hooked onto linguistically, whilst the distance between signifier ('S') and signified (a little bit of the real ('*s*'), unsettled and arbitrary in itself, but established for the time being, in this combinatory signifying process) is maintained by the function of the imaginary. Each instance is thus the signifierisation $\frac{S}{s}$, the key to the tablature of which is: 'S' = the signifier (in the Symbolic); '–' = the bar; connector/divider of the Imaginary (autonomising function); '*s*' = the signified, as an ever-receding little bit of real, unsolidified and constantly (re)fading as it is approached.

On the infinity of the rings, it becomes clear here that it is only the Real that is truly infinite, in its homonomy. As symbolised—or signifierised—as 'bits' of it may become, it still shifts indescribably and incalculably beneath that process. Thus, linguistically, if—as Ferdinand de Saussure pointed out—language is differences without positive terms, the Real is this without differences as well. Likewise, in terms of topographical placement; as Koyré clarifies in *From the Closed World to the Infinite Universe*:

> The category or question: "where?" In infinity it has no meaning. The infinite is not something, a sphere, of which the center is everywhere and the limits nowhere; it is something of which the center is nowhere also, something that has neither limits nor center, something in respect to which the question "where?" cannot be asked, as in respect to it everywhere is nowhere, *nullibi*.[13]

Thus, whatever can become known of the Real—to whatever extent—once it has become so, it is no longer of the Real, but resides in or through the other orders.[14] Therefore, it must also be borne in mind that the topological plane of the Real is only ever—and can only ever be apprehended as—a useful indication, as opposed to anything like a faithful representation. As infinitised, the Real having no 'where' thus implies that at whatever point a signifierisation occurs—at whatever point a signified comes to be established (through a signifier—or what we could call *a* symbolic—imaginarily linking to the Real)—it will strictly be an *arbitrary* point (in *real* terms, that is).

* * *

From these presentations we can perhaps align major cuts with *holes*, and minor cuts with *displacements*. To go back to the cuts of *2001* that we above placed under these two headings (major and minor), we might read that the monolithic cuts present us with holes in that, taken respectively: the appearance of the monolith on the Earth of four million years ago creates a hole in the *Umwelt* of the apes (the world as it is experienced by them). Prior to it they were not properly *subjects*—certainly not *modern* subjects, to use the Kojèvean distinction—(they were existents with a type of sentience more akin to that of the herbivorous animal), and only thereafter—after its cut is made—do they become so: afterwards, homogeneous animality is separated out; the tapirs become seeable as food, and the means by which to make them into such become comprehendible (thereafter, the advancements of this initial science, or *technê*—including the transition into language—are thus only extensions of its initial violence, or aggressivity). It is the monolith's removal (its absence, its lack) that leaves a hole: when it is present, it is all-encompassing, smotheringly mesmerising; the apes cannot leave it alone: there is in fact a *whole*. In its going, there is a *hole*, which is endlessly tried to be refilled. Thus, what we subjectively encounter as a *whole* is only so because of a *hole*; or, in a word, '(w)hole', as Jacques-Alain Miller once put it.

In the second instance of the monolith's appearance—its exhumation on the Moon—a similar cut occurs, though this time within the realm of Artificial Intelligence (AI). In the film, prior to HAL, all the computers that had been seen functioned in accordance with their programming by humans, whether via pre-recorded analogue messages dictating instructions (the automated video of a female staff member informing visitors at the point of entry to the Clavius base of the rules concerning the voice-print identification security system), or digital apparatus (the astronauts' premonitory touchscreen tablets, or the navigation controls of the film's various spacecraft, the cockpits of which are seen: the *Orion III* commercial flyer, which transports Floyd to *Space Station V*, and the *Aries 1B*, which transports him from there to the Moon, both run by Pan American World Airways; the Moonbus, which shuttles passengers from site to site on the Moon; and of course the *Discovery One*, and its EVA (Extra-Vehicular Activity) spacepods). HAL, however, *can think*; 'he' is a sentient being, and has a great deal of intelligence, as artificial, or *humanmade*, as it is. A constituent of intelligence, however, is perhaps its ability to escape or circumvent its (previous) confines; indeed, those of its creation. If this ability was artificially implanted in HAL, by humans, its escape or circumvention of the boundaries of this implantation puts it *beyond* the human; that is, those intelligent extensions are beyond the humanmade.

Of course, the HAL9000 computer was created in 1992 (12 January, to be exact), and thus preexists the cut brought about by the Moon's monolith; however, it is only after the cut—the screech of which turns out to be a signal directed at Jupiter—that HAL develops what could be termed an unconscious. His scrupulous dedication to—and 'greatest enthusiasm and confidence in'—the subsequent mission to the giant planet, and his devotion to the crew—which becomes murderous in its hystericisation—conflict to such an extent that in his wonderful store of knowledge something like a hole is created, which no code can fill: an entanglement occurs that cannot be resolved by the computer's preloaded logic. Its failure to cope is thus its *becoming-subject*. In this instantiation, then, the cut brings artificial intelligence up to speed with human, but takes it no further; that is, its seemingly limitless knowledge—which is in fact transcendentally limited—and seemingly infinite computing and calculating speeds are minor cuts in its trajectory—that displace former capacities and velocities—compared to the major cut of forming an unconscious. (After this point, any further incarnation of HAL's, and AI at large's, potentiality is inhibited in the film by Bowman's destruction—which takes place through a prolonged process of desubjectivation—of HAL from within his 'nerve-centre'. Clarke, however, provides such instances in the book's sequels.)

In its third transformative manifestation, the cut that the monolith brings about is into space and time themselves; it is an incision into the fabric of the space-time continuum. Like Odysseus in the environs of

Aeaeae, Bowman, in his spacepod, is transported—in the most ethereal sense of the word—through an extraordinary space odyssey, in which impossible shapes form; colours and contours mix in beyond-otherworldly sprawls and sprees; planetary landscapes appear in utterly unrecognisable hues and unreal dimensions; and the borders between inside and outside, unconscious and universal (in the sense, that is, of the universe), id and ego (the superego having seemingly dissolved), the seeing and the seen, sanity and madness, etc., collapse. If Bowman is not yet beyond the form of human subjectivity that belongs to the epoch prior to his entering the Stargate, it is as if the dimension in which the fear or anxiety he might feel in this situation has been cut so radically that these affects cannot affect him here; he seems only to be a spectator, hollow and frozen, and the affects themselves seem to be without their former animating dimensions; lost in space, or rather, lost in a hole in space. And the horizon of this event was of course a black hole: the monolith itself.

Lastly; appearing at the end of the bed in the eerie 'hotel' room after Bowman's rapidly cut-up cycling through the ageing process, the monolith hails the dawn of a beyond of subjectivity, as we know it, itself. Levitating above the bed, the enigmatic so-called Star Child retains human features—it is foetus-like, but with wide-open, very large eyes (the opposite of *eyes wide shut*)—and thus is not an alien species, or the overcoming of humanity *in toto*, but something that, through combination with the monolithic force (whatever it is), is subjectively new. In the last shot, in a sort of aura-womb, the Star Child floats in space off of Earth and turns its eyes towards the planet. In this scene, we are not given any temporal coordinates—the age at which the Earth is is precisely unknown here (no indicators can be seen beneath its clouds, nor are there any humanmade satellites visible)—and we are thus left with the question (reinforced by the repetition of Richard Strauss' *Also Sprach Zarathustra* at this point) whether this is the fulfilling of Friedrich Nietzsche's prophecy of the *Übermensch*, or of that of the eternal return of the same (that is, as a cut that enacts a sort of inescapable rewind to the beginning of the movie, and the (re)playing-out of its events). The ending is nothing if not ambiguous, and the ambivalence of this dichotomy can certainly be read into its open-endedness.

In 'The Subversion of the Subject and the Dialectic of Desire', Lacan highlights how through psychoanalysis the structure of the subject can be subverted, how it can come to *escape*—however briefly—perhaps into an interstice that cannot support it; that is, of the real, or of jouissance. It must return, but it returns with a modulation, perhaps an affirmation, in the Nietzschean sense: indeed, if we could never bring anything in from the cold of the real, we would have no warmth at all. Thus, in its major cuts, the meaningful—all too meaningful—economy of signifier-signified—the signifying chain that animates the subject of the unconscious—can be subverted by centralising the holes of analysis' discourse or dialectic. Indeed, as Lacan states: 'if linguistics enables us to see the signifier as the determinant of the signified, analysis reveals the truth of this relationship by making holes in meaning the determinants of its discourse.'[15] If such holes can of course not be subsisted in, what is modulated by them perhaps partakes of Lacan's neologism '*jouis-sens*', known in English as 'enjoy-meant' or 'enjoy-meaning'; that is, what is brought back from the jouissance of the hole is something that can combine in meaning, and form a modulation, which is perhaps to say that when a major cut can be truly glimpsed a minor cut might be made within us (that is, as a major cut truly made would be a sublation so revolutionary as to render it imperceivable or inconceivable from our current subjective state, or in the current order of things). (An extension of the above 'general' and 'processual' presentations might reveal clinical workings of these cuts, which will be returned to momentarily.)

In *2001*'s (non-exhaustively) above-listed 'minor cuts' we find displacements; first, and most obviously, in the famous jump-cut, where continuation, of science and violence (which are given something of a co-dependent origination), is seen to be the true operation of four million years of human evolution; that is, over complete change, or transcendence. In close conjunction; throughout Koyré's historicity, lines of continuity are highlighted as running underneath and connecting the scientific revolutions of the modern *epistēmē*, an example of which can be found in his analysis of Galileo Galilei's telescope, which operates a form of displacement on the visible, but does not transcend visibility, or the operation of seeing (that is, it does not infinitise or go beyond the infinite). As he states: 'the telescope does not change the situation. It allows us to see more stars than we did before its invention; it enables us to transcend the *factual* limitation of our sense of seeing; but it does not remove its essential structure. With as without the telescope, things at an infinite distance cannot be seen. The optical world is finite.'[16] Lacan was also always quick to stress the sense of *continuing revolving* in the 'Copernican revolution'—that is, its minor cut in highlighting the heliocentrism of our solar system—although it is anticipatory of a major cut in its decentring of human subjectivity, which can thereafter no longer be *scientifically* put at the centre of the universe.[17] Indeed, in this sense, the Darwinian theory of evolution represents something of a culmination in scientific revolution alongside an inherent realisation of the structure of its historicity.

Just as prior to the jump-cut from bone to spacecraft was the precedent of the monolithic cut, before the transition to the Jupiter mission was the cut instantiated by the monolith and its screech on the Moon that manifests itself in HAL's malfunctioning, or, indeed, *functioning in accord with the unconscious* (it is indeed only from the monolith's 'behaviour' on the Moon that the mission comes about, and that HAL is put onboard the ship). Here, the limits of previous space exploration are displaced as further reaches of the solar system can become available to travel to (the practicalities of which are elucidated in detail in Clarke's account, and meticulously, silently reproduced in Kubrick's).

The film's intermission—which is a cut in the sequence of the Jupiter mission; that is, precisely *inter-mission*—is preceded by a HAL's-eye-view shot of Bowman and Poole discussing HAL's erratic behaviour in a hermetically sealed spacepod, which is thus conveniently soundproofed, as the astronauts do not want HAL to hear their talk. However, HAL can see through the spacepod's window with the glowing red 'eye' of his control panel, and it becomes apparent that he is able to lipread, an advancement in computer recognition technology displacing what was thought possible of it at the time of the film's making. Indeed, Clarke stated in an interview: 'the one episode in the film which I thought improbable—and this was Stanley's idea, *not mine*—was HAL lipreading. Well, now they are training computers to lipread. So, Stanley was right, and I was wrong.'[18] (It was in fact Lockwood who came up with the suggestion of HAL's lipreading during the film's production.[19]) Yet, it was not this advancement in computerised processing that had propelled AI in the movie into a new state of being entirely; it was thus only an ultimately realisable extension of a computer's programmable abilities.

Finally, in the cuts of death itself as they appear in the film—most strikingly in that which is the result *of* a cut: the cutting of Poole's oxygen supply—the most radical subjective cessation is put in play. Indeed, as far as human subjectivity is (currently) concerned, death is the most radical of all cuts, in its finitude, but whereas it might seem like a *hole* that is created through it, it should more properly be considered a *whole*, due to the complete cessation of the conditions for and coordinates of life, which themselves encircle a hole; that of the possibility of death; that created by our subjectivity as *beings-towards-death.*[20] Indeed, the conditions for and coordinates of life enact a circulation *as the death-drive*. What is thus displaced, in death, is the death-drive itself (that is, from a subjective singularity as its embodiment; i.e., it gets *displaced* back into life *of others*).

In sum; when Lacan states in 'The Subversion of the Subject' that 'starting with Freud, the unconscious becomes a chain of signifiers that repeats and insists somewhere (on another stage or in a different scene, as he wrote), interfering in the cuts offered it by actual discourse and the cogitation it informs', he is highlighting that 'discourse' (that realm of the social, of *link*) offers up to the subject (of the unconscious) its major cuts—in the form of holes—the displacements of which (*by* the unconscious) bring about the animating minor cuts through which the subject subsists.[21] However, it is when these unconscious signifiers begin to entangle, and knot together messily, that the minor cuts are elided, and symptoms—and their repetitions and reinscriptions—form.

* * *

In the following part of this chapter—which will form something of an intermission—we will work towards a theory of *symptom-cutting*, and lay it out in relation to the cuts involved in Lacan's clinical idea of the variable-length session (and thereafter to those of Kubrick's unique style of film-editing).

In psychoanalysis, the unconscious is to be taken with the pinch of salt that it rubs into its wound, and this, so often, is the *symptom*. Drawing on the above topological work of the 'general' and 'processual' presentations, we might now set out here a heuristic display of *symptomal topologies*. Doing so is merely intended as a utilisation and theoretical extension of Lacan's late topological work in a fashion that can be utilised and extended, reworked, or completely restructured, in itself; which is to say that it is not to be taken doctrinally or dogmatically, but hopefully productively, clinically and critically. Structurally derived from the foundations of Freud's psychoanalytic theory and method, in Lacanianism there are three prominent psychical (what get inadequately called 'nosographical') categories (the 'Freudian structures', as Lacan referred to them): neurosis, psychosis, and perversion.[22] Within these structures there are also what Freud called 'dialects': hysteria and obsessionality in neurosis, and perhaps what has been extrapolated by Miller from Lacan's work on Joyce as the functioning of 'ordinary psychosis', for example. In terms of the symptomal processes involved in these structures, from the schematism of the 'processual' presentation can thus be derived ways of thinking these as links. In so doing, divisions of the Symbolic and Imaginary into discrete elements will be necessitated: those of the signifying chain and its signifiers, in the Symbolic, and the autonomising function(s), and its combinatory signified(s), in the Imaginary, and, in each *individual* instance—that is, in each processual instance of an individual unconscious—the Real should be taken as a ring (whilst this ring is *at the same time* still part of the infinite line of the Real, of course), although it still should be seen as the key ring, which, if it were to be cut, would dissolve the whole structure, as will be in demonstration throughout.

Thus, initially, we might make the distinction that neurotic (and perverse; *inversely* to neurotic) links are 'Borromean' in their properties, whereas psychotic ones are not (and thus function in another manner, or might necessitate a form of *sinthomic* 'repair').

To set these out: for neurosis, we might depict the hysteric link as in Fig. 3.9 (that is, as a replica of the third stage of the 'processual' presentation of the homonomy-heteronomy-autonomy intersectionality).

We can highlight here that it is the threading of the Symbolic's signifiers through the Real and Imaginary that constitutes the neurotic's signifieds (this will be different in the case of psychosis). In hysteria, these signifieds thus subsist discretely from one another (as 1, 2, 3, 4, 5, etc.), and if a signifier were to become hystericised, and thus (its signified become) symptomatic, it would be this particular symptom-formation that could be worked on in analysis, and its signifier, or its imaginary tie, that could be aimed at being cut, to free the symptom from the linked structure and 'resolve' it. (Another signifier could of course then be retied to take its place, which could itself then become hystericised, metonymically, etc. There is no limit to the signifier-signifieds in or on the link; any quantity would only indicate the 'vocabulary' of the unconscious' structure (which is like a language); a 'vocabulary' which is thus open to thought processes on the one hand, and symptom-formation on the other).

For obsessionality, the weave can perhaps be presented slightly differently, although by still retaining the Borromean property. Instead of discrete autonomising functions of the Imaginary, we can portray the Imaginary as a single ring, which holds signifiers (of the Symbolic's signifying chain) in place against the Real, via a certain weaving, or braiding (which is to say; if the obsessional signifiers were all to be under the same weave, they would clump together). Thus, we can show them as one signifier 'behind' the Real, one 'in front', etc., yet all linked in Borromean fashion; although this is ultimately dependent, in terms of crossing points alone, on which way the Real initially goes to tie into a ring (which is not the case

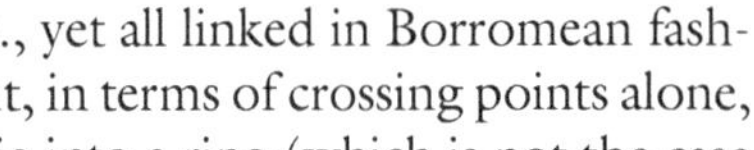

Fig. 3.9 Link of hysterical neurosis. Source: the author

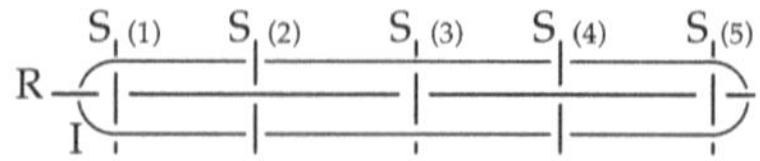

Fig. 3.10 Link of obsessional neurosis (the inversion of which is the link of perversion). Source: the author

if it remains infinite). What this means is that certain signifiers would 'slip off' of the Imaginary if not held in place by others before them.

Take the instances of 1 and 2 in the above diagram of the obsessional link (Fig. 3.10). 2 could not come away of itself from the link, unless 1 had already been removed. 1 holds 2 in place between itself and 3, but if it were removed, 2 could 'slip off' from the Imaginary via the gap it would have left, whilst 3 'reties' the link, in keeping 4 knotted in between itself and 5 (*ad infinitum*). (This of course also works *in reverse.*) Clinically, obsessionals often display the ability to talk (to 'do the work' of analysis), often of things that would appear quite 'objectively troubling', although these might not be the symptom. That is, there might be a tendency to concentrate on, or *compulsively repeat*, signifier 1, whilst it might in fact be signifier 2 that is at stake, that has become symptomatic. Signifier 1 would then need to be cut to allow signifier 2 to come to the fore, and begin the process of slipping away from the Imaginary. A clinical example might take the form of an analysand whose compulsive behaviour away from analysis involves an addiction-like repetition of an inhibiting action (say, online gambling; inhibitive perhaps of money accumulation, or social interaction beyond the computer and the digital green felt of poker and roulette table graphics). However, in analysis the obsessional might find themselves unable to talk about this, and thus talk instead about (what they may consider to be the *expected*) tropes of Oedipal relations; 'daddy-mummy-me', as Gilles Deleuze and Félix Guattari have it, etc. However, relations to—or between—these Oedipal players might one day change (say, if someone in the Oedipal triad is stricken with a grave illness); this perhaps in itself may cut that signifier—or ready it to be cut in the analysis (it might resonate with something in it: death, perhaps; a favourite theme of the obsessional)—freeing up the signifier 'behind' it, symptomatic of the compulsive behaviour. As a cut in the life of the analysand, this process might be phrased by them as: 'in relation to this major event, my symptom seems smaller-scale, and therefore I can bring it up now (that is, under this pretence).' (What is unavoidably implied in this topologisation is a sequentialisation of signifiers/symptoms, which of course proves problematic, but perhaps also representative of the problematics the treatment of obsessionality can face. What this sequentialisation thus seems to imply is that it might be more than one 'barring' signifier that needs to be worked on to free up a symptom—and, necessarily, other (perhaps non-symptomatic) signifiers, in the 'vocabulary'—to be able to assume the position of 'detachability'.) What thus links the neurotic structures, in this theorisation, is their *Borromeanicity*; yet it is the particular manifestation of this that separates them from each other.

In the *Three Essays on the Theory of Sexuality*, Freud put forward the thesis that '*neurosis is, so to speak, the negative of perversion*.'[23] However, this is in no way incompatible with Lacan's identifying perversion as a purely different structure to neurosis. It is. Yet it is also a structure which can be presented as partaking of the Borromean property. Indeed, as Dylan Evans helpfully elucidates:

> Freud's remark that 'the neuroses are the negative of the perversions' has sometimes been interpreted as meaning that perversion is simply the direct expression of a natural instinct which is repressed in [neurosis]. However, Lacan rejects this interpretation entirely (S4, 113, 250). Firstly, the drive is not to be conceived of as a natural instinct which could be discharged in a direct way; it has no zero degree of satisfaction. Secondly, as is clear from the above remarks, the pervert's relation to the drive is just as complex and elaborated as that of the neurotic. From the point of view of genetic development, perversion is at the same level as neurosis; both have reached the third 'time' of the Oedipus complex (S4, 251). Perversion therefore 'presents the same dimensional richness as [a neurosis], the same abundance, the same rhythms, the same stages' (S4, 113). It is therefore necessary to interpret Freud's remark in another way: perversion is structured in an inverse way to neurosis, but is equally structured (S4, 251).[24]

Thus, for comprehending perversion's structure against the above neurotic links, we can likely come closest to it if we were to swap the role of the signifiers of the Symbolic ('S's) with that of the Imaginary loop ('I') in the obsessional neurotic presentation (thus arriving at the inverse of that structure). Evans suggests—drawing on and condensing a schema of Lacan's in the *écrit* 'Kant with Sade'—that if the formula for (neurotic) fantasy is '$\$ \diamond a$', for perversion it becomes '$a \diamond \$$'; that is, fantasy's inverse.[25] This is because the pervert locates themselves—imaginarily—as locus of the

Other's enjoyment, or as the object through which the Other enjoys. In other words, in perversion the object *a* is identified with, not sought; its being sought—by the Other, '$\not{A}$'(which gets diluted into *others*; other people: '$\not{S}$'s)—is the 'turn-on', so to speak, and to (attempt to) fulfil the Other's desire—*completely*, at whatever symbolically destituting cost—the goal. Thus, as an inversion of neurotic fantasy, the moment of identification (of the imaginary) in perversion is with the *a* (which, as a notation, doesn't need barring—like the upper case 'A' of the big Other—as it is constitutively barred in its very structurality, as an always already missing or lost object; which is why the upper case 'A' does get barred, in its *ex-plicit* reflection of this quality in the *a*). It is in the barred Other ('$\not{A}$') that the pervert searches for a gap to fill; that is, what are seen as cuts in the Symbolic Other—or, to use the schematic notation: as '/'s (which bar 'A' as '$\not{A}$')—the pervert tries to imaginarily stop up, *as a*; that is, to complete (the jouissance of) the Symbolic (as *jouis-sens*).[26] Thus; whereas in the obsessional link, the symptomal signifieds are made up of individual symbolic signifiers being imaginarily linked to the real, in the perverse link, it is through individual imaginary identifications that the signifieds are produced. That is, '$S_{(1)}$', '$S_{(2)}$', etc. would get replaced with '$I(a)_{(1)}$', '$I(a)_{(2)}$', etc., which stand for so many positions (1, 2, 3, ...) of the object *a* that get identified with by discrete functors of the Imaginary (its autonomising functions; though objectally directed in this instance, as opposed to subjectively, as in the hysteric's link).

Yet, in the perversion formula derived from Lacan, it seems that the sights are set on the position of the barred subject; that is, *from* the *a* ($\not{S}$ being what lies beyond the lozenge's 'desire for', as Lacan labels it in 'Kant with Sade'[27]). That is, there seems here to be an unconscious idea that a form of subjectivity is arrivable at in the full realisation of (one's) objectality. To return to Miller's diagram of the surface fragment ('*a*') removed from the rectangle—excerpted in the 'Overture' (in Fig. 1.1)—in which the hole of the barred subject is left: *as* this hole, it is the *a* that is sought. For the pervert (or, more strictly, in the inverted fantasy of perversion), *as* the *a*, it is the *hole* that is sought (even if disavowedly; as in fetishism, which Lacan called the 'perversion of perversions'[28]). Thus, to further the above; in the inversion of the obsessional link, it is discrete moments of the Imaginary—*identifications*—that are linked by the Symbolic loop to the Real, forming the signifieds of perversion, and it is thus these moments of identification that can become symptomatic, liable to result in actings-out, or more consequential *passages à l'acte*; enactings of the object *a*, although ones in which the assumed mediation, or commands, of the Other may have become skewed in the pervert's purview.

In Seminar X, Lacan gives the example of Freud's case of a young homosexual woman—identified as one pertaining to the perverse structure—in which, in 'fling[ing] herself, *niederkommt*, let[ting] herself drop' onto a railway track, in a botched suicide attempt, she 'is brought into relation with what [s]he is as *a*', due to an impasse that does not allow her reaction to a sudden traumatic manifestation to be neurotically symbolised; that is, to take the form of a *linguistic* symptom.[29] Thus, it can be suggested that in the neurotic, the symbolic's passage to the real is via the imaginary; in the pervert, it is the imaginary that attempts to make a passage to the real, via the symbolic. If the symbolic is the mediator in the latter instance, then it is through it that the imaginary must pass; the symbolic thus takes on the attribute of an *outside* language—as opposed to the private language of a neurotic, which they attempt to make understood to or by the outside—into which an act, an object, can be written: the *falling* of a woman considered *fallen*.

It might be remarked that whilst a linguistic structure always underwrites the unconscious, its symptomal manifestations range, between the Freudian psychical categories. Thus, in neurosis symptoms manifest most prominently in language as it is thought and spoken (therefore, as the very stuff of the classical 'talking cure'); that is, as disturbances of the symbolic. In perversion, they manifest most prominently in acts—often *beyond the social* (that is, the symbolically strictured and structured; whether in statuted law, or morality *à la mode*)—in reaction-formation to *unsymbolisablity*; that is, when an *impasse* of symbolisation creates a *passage* to the (bodily) act. These may thus be construed as disturbances of the imaginary (considered the locus of the body, from the mirror-stage, and its misrecognitions, onwards).

In psychosis, it is perception that is affected symptomally. In terms of the topologies attributed to psychosis (by Lacan himself, and others), this is due to faults in the (*percipiens*') tie to *percepti*, and in the structure of the *percepti* themselves, within the psychotic link. That is, generally, what is perceived (*percepti*) is made up of the Real and Symbolic, and tied to (in the phenomenon of perception) by the Imaginary (of the *percipiens*, or the perceiving). In psychosis, however, it appears that the Real and Symbolic are directly tied together (not connected only through the Imaginary, as in a Borromean link), and the Imaginary insufficiently connected to them, in one way or another. That is, when describing a psychical structure derived from his knowledge of the work and life of James Joyce in Seminar

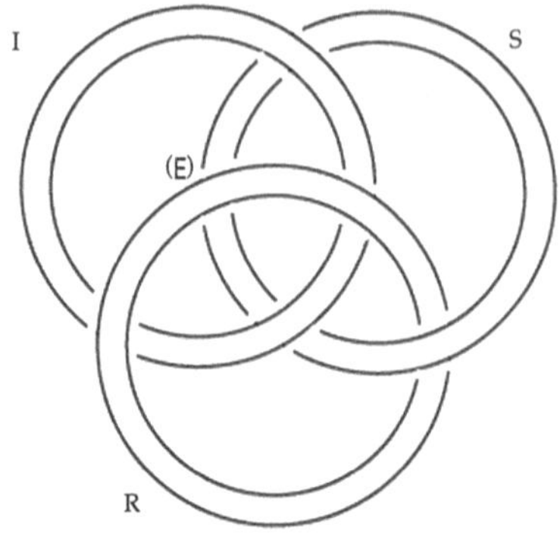

Fig. 3.11 Imaginary 'wedged' between linked Symbolic and Real. Source: the author

XXIII—known as *The Sinthome*—Lacan formulates a link in which, with Real and Symbolic directly tied, the Imaginary is 'wedged' between them (Fig. 3.11).[30]

As can be seen, the Symbolic ring ('S') threads directly through that of the Real ('R'), the Imaginary ('I') being held in place between these two. This formulation of 'Joyce's link' has since led to the designation of the functioning of 'ordinary psychosis', which—because the Imaginary is maintained in its position, unless an excess of pressure is applied to it—can typically present as a neurotic structure (or as its inverse), until a rupture (in which the escape of the bodily—and its perception—from the subject of this structure might be discerned). (Thus, prior to any rupture, the link may be misrecognised as Borromean, whilst it is in fact not.) To 'rectify' this possibility inherent in this link, a *sinthome* can be tied around the rings of the Real and Symbolic in the space demarcated '(E)' above—that is, that awaiting the *sinthome* (the siglum for which was chosen from Joyce's last major work *Finnegans Wake* in my previous work on Joyce and Lacan)—by weaving a ring from point '(E)' over the Symbolic, under the Real, back over the Symbolic, and again under the Real, and connecting it up as a full ring back at point '(E)' (Fig. 3.12).

To the function of the *sinthome* we will return after elucidating the last of the symptomal links that will here feature; that of psychosis (which is also repairable in a *sinthomic* fashion, although more fragilely). Lacan does not give a topology of psychosis in his late work (nor does he identify the Joycean structure as psychotic), but one fashioning of its possible construction is extrapolated by Raul Moncayo in his reading companion to Seminar XXIII, in which the Imaginary, instead

Fig. 3.12 The *sinthome* repairing link of Fig. 3.11. Source: the author.

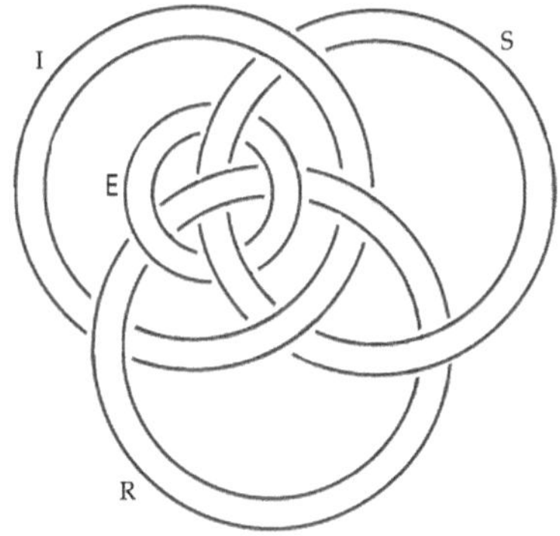

of being wedged between the linked Symbolic and Real, is completely separate to them, 'resting' on top of or underneath these linked rings.[31] A specially threaded *sinthome* is then able to tie the Imaginary ring back to the others. Prior to this, however, the Imaginary is in a sense out on its own, sometimes touching the linked Symbolic and Real (in which, due to their inextricable tie—as Miller puts it in relation to schizophrenia—there is no 'defend[ing one]self from the real by language, because[, here,] the symbolic *is real*), but not properly linking to them, and thus receiving what could be termed their echoes and mirages; thus *perceiving their influence*.[32] Moncayo's links—of, in his words, 'three psychosis' ('R', 'S', 'I') and 'four psychosis' (the three, retied with the *sinthome*: 'E'), brought in line with the ordering of the above Joycean links—are thus those of Figs. 3.13 and 3.14.

In this, the Imaginary 'rests' up against two points on the Symbolic and one on the Real, although it is in no way secured to these orders. With the ring of the *sinthome* fixed in place, a securement occurs, stabilising psychotic perception to an extent (although the Imaginary's room for wayward manoeuvre is not as constrained as in the link of 'ordinary psychosis' tied by the *sinthome*).

To return to symptoms and *sinthomes*: a clinical example of a psychotic symptom may be seen in an analysand who, in their speech, returns to an instance in their youth in which they had violently attacked a parent. They now hear voices, which are described as giving them a good hiding every now and then. This physical attribution to the voices of bodily attacking the analysand represents a displacement of the earlier event, through a metaphor that manifests itself in the perception (in itself displacing the

Fig. 3.13 Link of 'three psychosis'

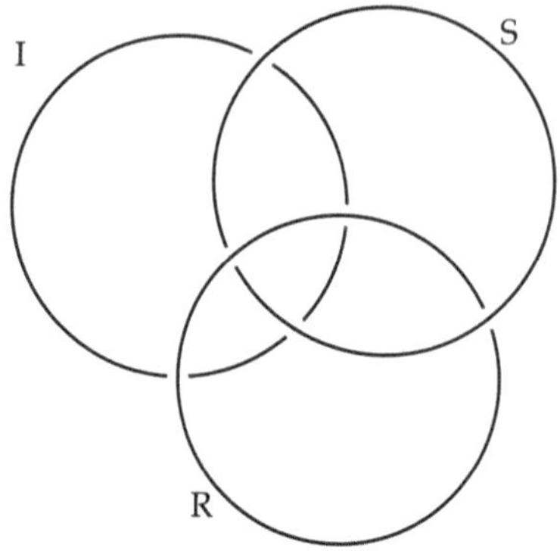

Fig. 3.14 Link of 'four psychosis'. Sources: the author

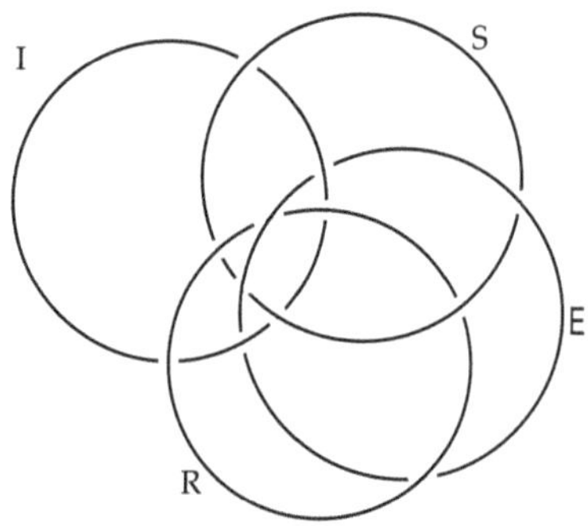

linguistic structure of the unconscious into its reappearance in the system of perception). In the structure of this psychotic symptom, a direct relation is foreclosed due to the fact that the Imaginary does not and cannot thread—or wedge itself—between the Symbolic and Real (that is, the symbolic is taken to be the real; it is perceived as such: *as* an actual 'hiding', not as the signifier 'hiding', from which its associations can propel interpretation and overdetermination. The signifier is thus literally hidden here; foreclosed).

However, the symbolic, and its signifying chain, can be modified in these versions of psychosis by the introduction of the *sinthome*. Thus, whilst it is not the signifier that founds the psychotic symptom (in this presentation of things), it could be deployed in various ways to adjust the psychotic experience. Thus, to adapt Moncayo's links to the plane of the signifying chain, we can show three symptomal stages of psychosis (Figs. 3.15–3.17):

1.)

Fig. 3.15 Link of psychosis on signifying plane. Source: the author

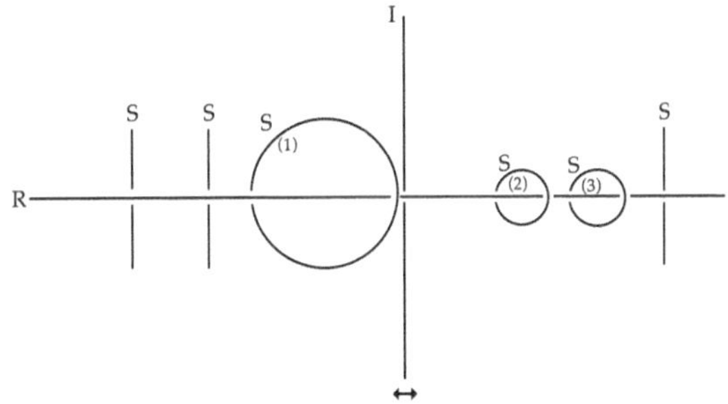

Thus, here we see the Real ('R') and Imaginary ('I') as infinite lines, and the Symbolic ('S') broken down into the signifying chain, between two signifiers of which (1 and 2; directly linked to the Real, as they are) the Imaginary finds itself (perhaps correlative to the two points of the Symbolic on which the Imaginary was seen to 'rest' above); its vacillation between these points—and in general (as it is untied to the linked Real and Symbolic)—is demarcated by the arrows. After the third signifier (unavailable to the Imaginary currently) the other signifiers of the chain are foreclosed and cannot be apprehended.

2.)

Fig. 3.16 Link of psychosis on signifying plane, with *sinthome*. Source: the author

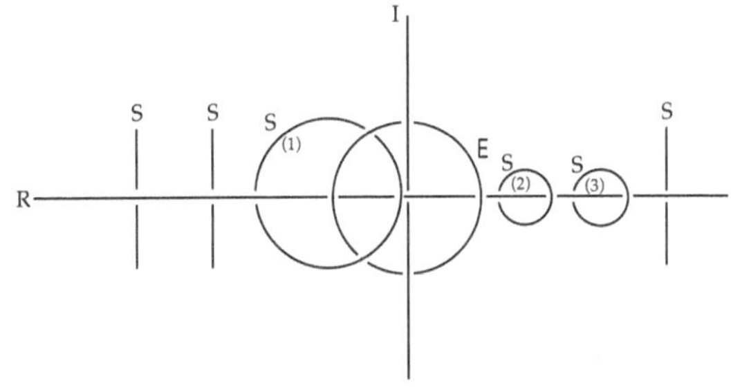

With the *sinthome* ('E') applied, the Imaginary gets tied in place, and is unable to escape the structure. It can no longer move as much between signifiers, although it has been given the safety net of the *sinthome* so that it does not simply slip away. '$S_{(2)}$', however, has become temporarily out of its reach—the others still foreclosed—until the next stage.

3.)

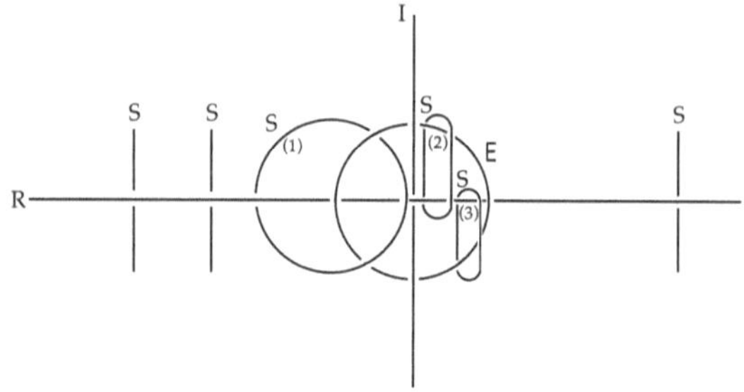

Fig. 3.17 Link of psychosis on signifying plane, with extended *sinthome*. Source: the author

In this third stage, the signifiers that might be available to the psychotic Imaginary can be linked to by extension of the *sinthome*; that is, to thread through '$S_{(2)}$' and '$S_{(3)}$', which previously wasn't available at all. These extra signifiers can thus hold the Imaginary more securely in place, whilst the others in the chain necessitously remain foreclosed, as per the structure; the non-Borromean structure of psychosis, that is. To couch this in clinical terms; if, classically, a signifier represents a subject to or for another signifier, in psychosis the subject seems to find themselves outside of this exchange. According to the above schemas, the psychotic imaginary has assumed the place of a signifier, occupying its position in the signifying chain, as if it were on the conveyor belt of language, the movement of which being completely beyond its control, beyond the reach of its autonomising function; that is, the imaginary here is utterly interpellated, and thus perceives the symbolic *as real*. Its sliding proximity to the two signifiers at its disposal puts their significations into the realm of perception; as hallucinations, for example. They are not connected to and separated from at once (in the manoeuvre of the unificatory/separatory principle) imaginarily, as occurs in the neuroses, and inversely in perverse identifications, but metastasise in various distortions of reality (brought about by a different—closer perhaps—relation to the real). In psychosis, meaning might *shine*; it might be transmitted by rays. The likelihood of the establishment of a transference in its analysis is diminished in that for psychosis *everything* might well already (be perceived to) *be* transference.

If we very simply imagine the Imaginary of the neuroses and perversion as a horizontal axis, and that of the psychoses as a vertical one, we can picture the imaginaries of analyst and analysand as 1.) linking (neurotic/pervert—neurotic/pervert)—and signifiers being put in exchange between them—but in the meeting of analyst and analysand when the psychoses are involved, imaginaries not connecting, but 2.) intersecting (neurotic/pervert—psychotic), or 3.) running parallel (psychotic—psychotic); that is, as:

1.) ——
2.) —|
3.) | |

The fourth relation—psychotic—neurotic/pervert: '|—'—(as well as the third) may have been what Lacan saw himself as ending up operating in his clinic, in identifying as psychotic himself in his late stages.[33] The cuts that this analytic set-up provoked could often be radical, as attestations of Lacan's late clinic bear out.[34]

What might be aimed at, then, in analysis of psychosis, is the sliding of signifiers towards the imaginary to fasten it in place more securely. Thus, in the treatment of its 'raw' state (versions of which—paranoia, schizophrenia, etc.—are omitted here), the imaginary's connection to the symbolic could be concentrated on, with its realisation as symbolic (that is, as not overwhelmingly *real*) as the goal thereof. This could perhaps be presented as the closing-in around the imaginary of '$S_{(1)}$' and '$S_{(2)}$' in stage 1 above (of course with an emphasis on the supportiveness of the closing-in, as opposed to its being taken in a persecutory sense). The establishment of a *sinthome*—which could be aimed at in analysis, through the *assumption* of the irreducible aspect of a symptom, as Lacan phrases it in his late work—would prevent the imaginary escaping (which might typically manifest in some form or other of an 'episode'), as seen in the second stage. In the third; in the treatment, the imaginary might be further buttressed by weaving the *sinthome* onward through the next two signifiers linked to the real ('$S_{(2)}$' and '$S_{(3)}$'), to secure the imaginary at two further points. In this structure, this thus seems to be the ultimate result of the last re-sort.

Finally, to return to the Joycean structure of psychality; ordinarily, with the imaginary wedged in place, symptoms typical of the neurotic structure

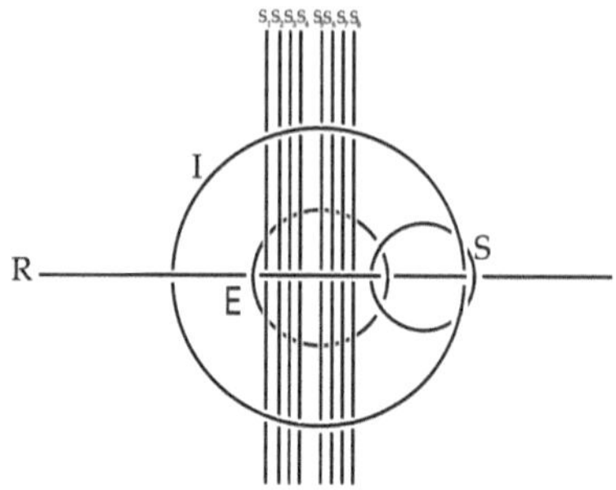

Fig. 3.18 'Joyce's link'. Source: the author

might present; that is, until some pressure forces a dehiscence of the imaginary, which the treatment could work on re-wedging. However, if the structure's *sinthomic* repair has been carried out (with(in) or without analysis; Joyce's was certainly of the latter, as Lacan is constantly at pains to remind us) then something new can occur with signifiers, in the realm of creation, or *creative writing*; that is, the signifier can come to *stuff the signified*, as Lacan said of Joyce's compositional practice in Seminar XX.[35] This can thus be presented as Fig. 3.18.

Here, it is as if signifiers themselves have broken loose of the signifying chain. Whilst the Symbolic order ('S') is still directly linked to the Real ('R'), signifiers ('S_1', 'S_2', 'S_3', etc.) are now able to thread through the reattached Imaginary ('I') and the *sinthome* ('E'), which function in conjunction as the place of the signified (stuffed with the signifier). Whilst signification thus abounds, the risk of its overloading the subject seems to have been countered by the securement enabled by the *sinthome*; that is, there can be a push in this particular linking made towards taking signification as far as it can go, cranking up its level of tension and intensity, its *intension*—to and perhaps beyond infinity, as the signifiers' presentation as infinite lines here suggests—whilst the Symbolic remains grounded, separating this adventure of sense from madness.[36] (It is as if the end of a language might have been reached, and its beyond thus become explorable; Joyce indeed said that he had taken English as far as it could go with *Ulysses*, the beyond of which was *Finnegans Wake*.[37] In *2001: A Space Odyssey*, the Stargate conducts a similar reaching-beyond of subjectivity, carried out in the film on the visual and spatio-temporal plane, as will be explored in the coming chapter).

Thus, with these presentations of links pertaining to the Freudian structures (in which untyings and retyings of themselves might always be taking place), we are enabled to envisage different ways in which, and means by which, symptoms can be cut, modulated, or sutured (in this last instance with the *sinthome*, which can come about through the *assumption* of a symptom—that is, connecting to the symptom's *real*—locatable for instance in the *assumption*, or accepting-as-true, of one's structure as psychotic, at any one time, from which sutural work with signifiers might commence). Specifically in terms of *symptom-cutting*, we might postulate that in the structures of neurosis—in which signifiers *flare* into symptoms—and that of perversion—in which certain identifications (with objects *a*) can become symptomatically dangerous or unhealthily detrimental to the analysand—their cuts can come about in analytic treatment through its cuts, major and minor: interpretations, silences, scansions, moments of concluding, highlightings of the holes in the analytic dialectic's discourse.

Retrospectively, these cuts may be seen as major or minor according to whether they had brought about a symptom's resolution or dissolution, or whether they had modified or modulated it in the direction of these results (obviously, if a cut had been attempted in the service of these results with neither achieved, it cannot be said to be either minor or major, and if a cut were made with inverse, unintended effects—unexpected retrogressions, triggerings, episodes, or outbreaks—they too could be adversely minor or major).

One practical method of cutting—that could result in either (or neither) of these results—that Lacan employed was the variable-length analytic session. In 'The Function and Field of Speech and Language in Psychoanalysis', Lacan broaches this subject, turning to discussion of the length of sessions.

> Here again it is a question of an element that manifestly belongs to reality, since it represents our work time, and viewed from this angle it falls within the purview of professional regulations that may be considered predominant.
>
> But its subjective impact is no less important—and, first of all, on the analyst. The taboo surrounding recent discussion of this element is sufficient proof that the analytic group's subjectivity is hardly liberated on this question; and the scrupulous, not to say obsessive, character that observing a

> standard takes on for some if not most analysts—a standard whose historical and geographical variations nevertheless seem to bother no one—is a clear sign of the existence of a problem that analysts are reluctant to broach because they realize to what extent it would entail questioning the analyst's function.[38]

What he is highlighting here is the somewhat arbitrary—or *geo-historically determined*—setting of the length of sessions at the '50-minute hour' mark (which does not correspond to the 'time' of the unconscious). Cutting the session earlier (this happened *in extremis* in Lacan's late clinical practice)—or even lengthening its moment to conclude beyond the standardised time (which happened far less, and to the idea of which Lacan was less inclined; although Winnicott would often operate in this way, for example)—could enable something to occur at a precise point, which the analysand could 'take away' with them. This not only includes a cut leaving the analysand with a certain signifier on which to ruminate, it could also entail a challenge being made to an analysand's mode of discourse, for example. That is to say, if the cut came suddenly, during a flabby moment of repetitious literary criticism conducted from the couch—as is Lacan's example in 'The Function and Field'—over the time between sessions this could be reflected on and questioned, and in the next session something more subjectively concerning might become the topic that speech alights on, which might not have been talked about otherwise.[39] Further, in anticipating the end of a session, an analysand has (a certain) time to prepare, perhaps by constructing a finale, conclusion, or summative statement which ties everything together (a bit too) neatly, or by watching the clock, specifically so as to put the blame on it when the time runs out that there was not enough time to get to the point, or to cover everything. Thus, Lacan states:

> The ending of a session cannot but be experienced by the subject as a punctuation of his progress. We know how he calculates the moment of its arrival in order to tie it to his own timetable, or even to his evasive maneuvers, and how he anticipates it by weighing it like a weapon and watching out for it as he would for a place of shelter.[40]

Indeed, it can be on the punctuation of an analytic session's scansion that so much can come to rest, for as Lacan says: 'punctuation, once inserted, establishes the meaning; changing the punctuation renews or upsets it; and incorrect punctuation distorts it.'[41] These risks taken in the dialectic of analysis are those that might determine whether or not a symptom will be given (new) meaning, or become further upset, or be distorted; the punctuation of its cuts might indeed have minor or major effects, re- or defamiliarising effects; and might become the moment in which the dehiscence of a troubled signifier or identification might be effected.

* * *

Kubrick in his film-editing practice would also introduce certain types of cut that would unsettle standardised methods in editing, epitomised in *2001* by what we could label the 'short cuts' to frozen frames of Bowman's contorted face—seen through the visor of his space helmet—during the Stargate sequence. These cuts are of a few frames in length; enough, that is, to constitute the effect of a 'glimpse' (whereas a single frame would likely be missed by a viewer, or would come to represent something only seen in a flash, which would thus not be distinguishable, as Tyler Durden (Brad Pitt) well knew in *Flight Club* (1999), in which he is seen splicing stills of a throbbing phallus into the film-reels of family pictures in his job as projectionist at a cinema.[42] Speaking somewhat to the psychoanalytic register, the dialogue that follows this revelation is: "Nobody knows that they saw it, but they did.' 'A nice, big cock." Durden himself—as unconsciously structured as he is—has already appeared onscreen in exactly this manner four times before his diegetic entrance. Also, of course, the repetition of 'I am Jack's [this and that]' echoes the usage of the proverb 'all work and no play makes Jack a dull boy' in Kubrick's *The Shining*). These types of cut reappear in *A Clockwork Orange* (in Alex's (Malcolm McDowell's) ecstatic bloody-fanged grin in his Beethoven-induced daydream of destruction), and most famously in *The Shining*, in the cuts to the Grady twins (and later their mangled bodies at the end of the corridor in the Overlook Hotel); to Danny's (Danny Lloyd's) Munchian scream; and to the door with 'Redrum' scrawled in lipstick on it. (Martin Scorsese could be said to operate the opposite to this—in the stretching out of scenes to force reconcentration on their subject matter—such as in the prolonged moment in *Taxi Driver* (1976) when Travis Bickle (Robert De Niro) drops an Alka-Seltzer into his drink and watches it fizzle). In the operation of these cuts, space and time distort, and the holes of discourse are drawn attention to.

Notes

1. Jacques Lacan, 'The Instance of the Letter in the Unconscious, or Reason Since Freud' [1957], in *Écrits*, p.417. Lacan states in 'The Subversion of the Subject' that 'we analysts must bring everything back to the cut qua function in discourse, the most significant being the cut that constitutes a bar between the signifier and the signified.' See Jacques Lacan, 'The Subversion of the Subject and the Dialectic of Desire in the Freudian Unconscious' [1960], in *Écrits*, p.678.
2. Jacques Lacan, 'On a Question Prior to Any Possible Treatment of Psychosis' [1955–1956], in *Écrits*, p.486, note 14. The R schema is depicted on p.462.
3. Ibid.
4. Ibid., pp.486–487, note 14.
5. For the theses pertaining to the unificatory/separatory principle, see Bristow, *Joyce and Lacan: Reading, Writing, and Psychoanalysis*. In a forthcoming review of this work, Raul Moncayo offers an interesting rephrasing of it as the 'conjunction/disjunction principle'.
6. Lacan, *Seminar IV: The Object Relation*, p.50. Indeed, to locate 'the first Freudian dialectic of the theory of sexuality', it might well be worth going back to the first edition of Freud's *Three Essays on Sexuality* of 1905, in which he states, of the object relation: 'the process of finding an object, prepared for since the earliest stages of childhood, takes place on the psychical side. At a time at which the first sexual satisfaction was still linked to the intake of food, the sexual drive had a sexual object outside the infant's own body—the mother's breast. It was only later that the drive lost that object, just at the time, perhaps, when the child was able to form an overall idea of the person to whom the organ affording it satisfaction belonged. As a rule, the sexual drive then becomes autoerotic, and only after the latency period has been surmounted is the original relation restored. It is not without good reason that the sucking of the child at the mother's breast has become the model of every loving relationship. The finding of an object is in fact a re-finding of it.' See Sigmund Freud, *Three Essays on the Theory of Sexuality: The 1905 Edition*, ed. by Philippe van Haute and Herman Westerink, trans. by Ulrike Kistner (London: Verso, 2016) p.73. In the *Standard Edition* translation—replicating the 1924 edition—James Strachey points out in a footnote that this 'paragraph[,] written in 1905, does not appear to harmonize with the remarks on the subject [...] written in 1915 and 1920.' See James Strachey, 'Footnote 1', in *Three Essays on the Theory of Sexuality*, in *SE*, VII, p.222.
7. Jean Laplanche, *New Directions for Psychoanalysis* [1987], trans. by David Macey (Oxford: Basil Blackwell, 1989) p.45. In 'Beyond the "Reality

Principle"', Lacan states: 'the psychoanalyst, in order not to detach analytic experience from the language of the situation that it implies, the situation of the interlocutor, comes upon the simple fact that language, prior to signifying something, signifies to someone.' See Jacques Lacan, 'Beyond the "Reality Principle"' [1936], in *Écrits*, p.66.

8. Jacqueline Rose, 'The Imaginary' [1975], in *Sexuality in the Field of Vision* [1986] (London: Verso Radical Thinkers, 2005) p.189. Rose is here quoting (and translating) from the anonymously-authored article 'Le clivage du sujet et son identification', *Scilicet*, 2/3 (1970) 103–136 (p.120). Unless written by Lacan, all articles in *Scilicet* were anonymous. Lacan himself utilises the topology of the cross-cap in his ninth Seminar, *Identification*.
9. In *Full Metal Jacket* there is a failure of a major cut, which is attempted on the part of Private Pyle (Vincent D'Onofrio), and which in the end culminates only in the completion of his 'major malfunction' (in Gunnery Sergeant Hartman's words). That is; followed by the extraordinary cut from the boot camp to 'the shit' of Vietnam, Private Pyle's *passage* to the acts of murder and suicide does not change the trajectory of the army's punishing training and its result in a catastrophic campaign of war. If Pyle has felt the systematic effects of this trajectory, the one thing his act does not change is this systematicity itself; it can only be seen by the others as an aberrant misfortune.
10. Lacan, 'The Subversion of the Subject and the Dialectic of Desire', in *Écrits*, p.678.
11. Jacques Lacan, 'The Situation of Psychoanalysis and the Training of Psychoanalysts in 1956', in *Écrits*, p.392.
12. It is Adrian Price who flags this up in a note to the English edition of Seminar XXIII. See Lacan, *Seminar XXIII: The Sinthome*, p.223, n.3.
13. Alexandre Koyré, *From the Closed World to the Infinite Universe* [1957] (Baltimore: The Johns Hopkins University Press, 1968) p.201.
14. As Lacan delineates of the Symbolic order in this regard in Seminar IV: 'th[e] signifying form gives a scansion to the operation of transformation which translates movement into substitution, the continuity of the real into the discontinuity of the symbolic.' See Lacan, *Seminar IV: The Object Relation*, p.410.
15. Lacan, 'The Subversion of the Subject and the Dialectic of Desire', in *Écrits*, p.678.
16. Koyré, *From the Closed World to the Infinite Universe*, p.84.
17. 'Indeed, Freud himself compared his discovery to the so-called Copernican revolution, emphasizing that what was at stake was once again the place man assigns himself at the center of a universe.' See Lacan, 'The Instance of the Letter in the Unconscious', in *Écrits*, p.429. Later—in 'The Subversion of the Subject'—he draws out his Copernican response much

more elaborately: 'We must home in more precisely on what Freud himself articulates in his doctrine as constituting a "Copernican" step.

For such a step to be constituted, is it enough that a privilege should be revoked—in this case, the one that put the earth in the central place? Man's subsequent destitution from an analogous place due to the triumph of the idea of evolution gives one the sense that such revocation implies an advantage that is confirmed by its constancy.

But can we be so sure this is an advantage or real progress? Does anything make it seem that the other truth, if we may so term revealed truth, has seriously suffered as a result? Don't we realize that, by exalting the center, heliocentrism is no less of a lure than seeing the earth as the center, and that the existence of the ecliptic probably provided a more stimulating model of our relations with truth, before it lost much of its interest when it was reduced to being no more than the earth bowing assent?

In any case, it is not because of Darwin that men believe themselves to be any the less the best among the creatures, for it is precisely of this that he convinces them.

The use of Copernicus' name as a reference has more hidden resources that touch specifically on what has already just slipped from my pen regarding our relation to the true—namely, the emergence of the ellipse as being not unworthy of the locus from which the so-called higher truths take their name. The revolution is no less important even though it concerns only "celestial revolutions."

From that point on, to dwell on it no longer means simply revoking some idiotic notion stemming from the religious tradition, which, as can be seen well enough, is none the worse for it, but rather of tying more closely together the regime of knowledge and the regime of truth.

For if Copernicus' work, as others have remarked before me, is not as Copernican as we think it is, it is because the doctrine of double truth continues to offer shelter to a knowledge that, up until then, it must be said, appeared to be quite content with that shelter.' See Lacan, 'The Subversion of the Subject and the Dialectic of Desire', in *Écrits*, p.674.

18. Arthur C. Clarke, in *2001: The Making of a Myth*, dir. by Paul Joyce, 2001.
19. Indeed, Patrick Webster prods his readers to take 'note that some of the most iconic moments in Kubrick's work appear to have derived from collaboration with actors. For example: Malcolm McDowell's singing during the attack on Mr. and Mrs. Alexander; Jack Nicholson's gleeful shout of 'Here's Johnny!'; Gary Lockwood's idea for HAL to lip-read Bowman and Poole's lips in the space-pod; Lee Ermey's improvised dialogue in *Full Metal Jacket*, and so on.' See Patrick Webster, *Love and Death in Kubrick: A Critical Study of the Films from* Lolita *through* Eyes Wide Shut (Jefferson, North Carolina: McFarland & Company, 2011) p.303, n.14.

20. Concerning this whole or absolute cessation, a (minor) criticism that could be levelled at Lars von Trier's apocalyptic *Melancholia* (2011) is that its last cut—to the film's credits—should have been to black and silence (the absence of image and sound, implying the absence of that to which they are visible and audible) only, so as to have rendered the world's end fully utter (and that it thus may have placed its credits at the start).
21. Lacan, 'The Subversion of the Subject and the Dialectic of Desire', in *Écrits*, p.676.
22. Lacan deploys the term 'Freudian structures', for example, in 'On a Question Prior to Any Possible Treatment of Psychosis', in *Écrits,* p.449. As Dylan Evans states, however, it is the term 'clinical structures' that seems to 'predominate [...] in the writings of Lacanian psychoanalysts today.' See Evans, 'Structure', in *An Introductory Dictionary of Lacanian Psychoanalysis*, p.196.
23. Freud, *Three Essays on the Theory of Sexuality: The 1905 Edition*, p.26.
24. Evans, 'Perversion', in *An Introductory Dictionary of Lacanian Psychoanalysis*, p.142. To Evans' references to the French of Seminar IV, compare Lacan, *Seminar IV: The Object Relation*, pp.113–114 and pp.289–290.
25. See Evans, *An Introductory Dictionary of Lacanian Psychoanalysis*, p.142, in which he refers to Jacques Lacan, 'Kant with Sade' [1963], in *Écrits*, p.653.
26. Thus, in her book on perversion, Stephanie Swales suggests: 'the perverse fundamental fantasy is better written $a \Diamond \not{A}$, or the object-cause of jouissance in relation to the Other. In contrast to the obsessive's negation of the Other, shown by the omission of the Other from $\not{S} \Diamond a$, this formula ($a \Diamond \not{A}$) emphasizes the pervert's necessity to prop up his relation to the Other. Furthermore, just as the obsessive does everything in his power to deny his own lack, such that his fundamental fantasy might be written $S \Diamond a$ instead of $\not{S} \Diamond a$, so too does the pervert make every effort to cancel out the lack in the Other, such that his fundamental fantasy might be written $a \Diamond A$ instead of $a \Diamond \not{A}$.' See Stephanie S. Swales, *Perversion: A Lacanian Psychoanalytic Approach to the Subject* (Hove: Routledge, 2012) p.93. The writing of these last formulae without the bars is thus to suggest their purely *fantasmatic* bearing.
27. See Lacan, 'Kant with Sade', in *Écrits*, p.653. Lacan here states: 'the lozenge $\Diamond$ is to be read as "desire for," being read right to left in the same way, introducing an identity that is based on an absolute non-reciprocity.' Thus, the non-reciprocity between $\not{S}$ and a is of the same order when their positions are reversed; its effect, however, of course the inverse (i.e., in perversion, to that of neurosis, confirming Freud).

28. Lacan, *Seminar IV: The Object Relation*, p.223.
29. Lacan, *Seminar X: Anxiety*, p.110.
30. See Lacan, *Seminar XXIII: The Sinthome*, p.130.
31. See Raul Moncayo, *Lalangue, Sinthome, Jouissance, and Nomination: A Reading Companion on Lacan's Seminar XXIII on the Sinthome* (London: Karnac, 2016) pp.v–viii. Moncayo's extrapolation of a link of neurosis (depicted as three unconnected rings), however, seems contrary to its workings in Seminar XXIII.
32. Jacques-Alain Miller, 'A Contribution of the Schizophrenic to the Psychoanalytic Clinic (Original title, '*La clinique d'ironie*') [1988], trans. and ed. by Ellie Ragland and Anne Pulis, *The Symptom: Online Journal for Lacan.com*, 2 (2002), http://www.lacan.com/contributionf.htm. Regarding the manifestation of the structure as schizophrenia, Miller here in effect stresses that because of the inextricably linked Symbolic and Real the Other's inexistence is perceived, and an *ironic* approach is therefore taken to the social. On the 'influencing machine' attributed to experiences in the schizophrenic condition, see 'On the Origin of the "Influencing Machine" in Schizophrenia' [1919], in Viktor Tausk, *Sexuality, War and Schizophrenia: Collected Psychoanalytic Papers*, ed. by Paul Roazen, trans. by Eric Mosbacher and others (New Brunswick: Transaction Publishers, 1991).
33. On this, for example, see Stijn Vanheule, *The Subject of Psychosis: A Lacanian Perspective* (Basingstoke: Palgrave Macmillan, 2011) p.163: 'Lacan described himself as psychotic, or at least as someone who tried to be so[.] In a lecture at Yale University he said the following: 'Psychosis is an attempt at rigor. In that sense, I would say that I am psychotic. I am psychotic for the sole reason that I always tried to be rigorous'[.] Later on, a student returned to this and asked Lacan if he was actually psychotic. His answer then was that he was not psychotic enough: 'If I were more psychotic, I would probably be a better analyst'.
34. See, for example, the chapter 'Psychoanalysis Reduced to Zero', in Elisabeth Roudinesco, *Jacques Lacan: An Outline of a Life and a History of a System of Thought* [1993], trans. by Barbara Bray (Cambridge: Polity Press, 1999) pp.385–398.
35. Jacques Lacan, *Seminar XX: Encore*, p.37.
36. On Joyce's work's relation to madness, from which he claimed it was separated by a 'transparent sheet', see Bristow, *Joyce and Lacan: Reading, Writing, and Psychoanalysis*, pp.39–64. The phrase '[senseless] adventure of sense' is Jean-Luc Nancy's, and is discussed in the aforementioned work on p.130.
37. Joyce is claimed to have remarked, in French: "*Je suis au bout de l'anglais* ['I'm at the end of English]',' [...] to August Suter, and [...] to another

friend, 'I have put the language to sleep.'" See Richard Ellmann, *James Joyce: New and Revised Edition* (Oxford: Oxford University Press, 1983) p.546.

38. Lacan, 'The Function and Field of Speech and Language in Psychoanalysis', in *Écrits*, p.257.
39. See Lacan, Ibid. p.259: 'In experimenting with what have been called my "short sessions," at a stage in my career that is now over [...] I was able to bring to light in a certain male subject fantasies of anal pregnancy, as well as a dream of its resolution by Cesarean section, in a time frame in which I would normally still have been listening to his speculations on Dostoyevsky's artistry.'
40. Ibid., p.258.
41. Ibid.
42. This moment in *Fight Club* chimes with the contemporary revelation of the insertion of two non-consecutive frames of a topless woman into a background window in Disney's *The Rescuers* (1977), which was discovered in 1999 when a child paused a VHS copy at exactly that point, causing the recall of around 3.5 million copies of the film. See David Mikkelson, 'Emotional Rescue', http://www.snopes.com/disney/films/rescuers.asp

CHAPTER 4

Space

Abstract Whilst the outer space in which *2001* is set is vast, and the clinical space of psychoanalysis is intimate, their interrelations are explored in this chapter, which begins by highlighting the play that Arthur Clarke seemed to be premonitorily making on *Alien*'s tagline, 'in space no one can hear you scream', in the novel of *2001*: in space 'the nearest psychiatrist [i]s the better part of a hundred million miles away.' Further dealing in incommensurables, the chapter explores both the threat of nuclear extinction, which was a major concern to both Kubrick and Lacan, and the spatiality of the unconscious, through Freud's picturing of it in *Civilisation and Its Discontents*. This chapter also brings Gilles Deleuze's ruminations on the cinema of Kubrick into dialogue with Lacan's on Edvard Munch's painting *The Scream*.

Keywords Space • Psychoanalysis • Dave Bowman • HAL • Gilles Deleuze • *The Scream*

Of the tagline to Ridley Scott's *Alien* (1979)—'in space no one can hear you scream'—Clarke gives a subtler premonitory version in *2001*: in space 'the nearest psychiatrist [i]s the better part of a hundred million miles away.'[1] Later on in the book, he remarks: 'like every man of his age, Poole took it for granted that he could talk instantly, to anyone on Earth,

D. Bristow, *2001: A Space Odyssey and Lacanian Psychoanalytic Theory*, The Palgrave Lacan Series,
https://doi.org/10.1007/978-3-319-69444-3_4

whenever he pleased. Now that this was no longer true, the psychological impact was profound. He had moved into a new dimension of remoteness, and almost all emotional links had been stretched beyond the yield point.'[2] As Clarke's series carries on, references to psychoanalysis and its spatial set-up continue to appear, and the practice is treated with a certain fondness, if a slightly nostalgic one. In *2010*, for example, Floyd (whose name has the ring of psychoanalysis' forefather's: Heywood Floyd/Sigmund Freud) is reminded of the televisual exposure he has had to the practice—which evidently Clarke had well predicted the concurrency of in the far-off year of 2010—whilst a somewhat heavy-handed 'electronic psychoanalysis' is being carried out on the revived HAL9000 computer:

> How often he had seen videodramas in which disturbed youngsters were straightened out by all-wise descendants of the legendary Sigmund Freud! Essentially the same story was being played out in the shadow of Jupiter.
>
> The electronic psychoanalysis had proceeded at a speed totally beyond human comprehension as repair and diagnostic programs flashed through Hal's circuits at billions of bits a second, pinpointing possible malfunctions and correcting them.[3]

Further, in *2061* Clarke inserts a whole segment entitled 'On the Couch', in which Floyd's position on a couch is image-matched by a colleague against the classical psychoanalytical set-up: 'though almost any seat was confortable at one-sixth of a gravity, Floyd had tilted his to the fully reclining position and had clasped his hands behind his head. Van der Berg suddenly recalled that this was the classic position of a patient, in the days of the old and still not entirely discredited Freudian analysis.'[4] (This concluding stance is parodied in *3001* in a communication made via a 'thoughtwriter': 'maybe should get psychoanalysed like in your time—never understood how that Fraudian—mean Freudian ha ha—nonsense lasted as long as it did', of which joke Clarke leaves the victor indeterminate in not narrativising the intentionality behind the Freudian slip made via this spontaneous-thought communication medium.[5])

Today, the space of psychoanalysis—particularly in Lacanian circles—has begun something of a journey beyond the couch technologically, spreading into the realms of telephony and digital videolinking. This partial transformation represents something of an 'electronic psychoanalysis' not quite predicted by Clarke in his reliance on the couch, perhaps due to the fact that, in Ian Parker's words:

> This apparatus [of the couch] has been sedimented in psychoanalytic culture. [...] What a 'couch' should be became an issue; in Brazil, for example, one of the evangelical churches noticed that there was a market niche for this kind of therapy and proposed legislation to regulate psychoanalysis which stipulated the size and dimensions of this equipment. [...] In Japan there was an early debate in the history of psychoanalysis over the difference that might be engineered if the analysand were sitting in a chair faced away from the analyst, on a low sofa or on a low couch as such, and much was made of the difference between these types of furniture[.]
>
> [Yet,] what is most important is that the analysand is speaking, but not speaking *directly* to another person.[6]

It is curious to perceive throughout Clarke's references a move away from the centrality of speech (as spatialised as the initial references to it are)—through purely impersonal technique (the diagnostic repair and correction of HAL is perhaps a stance too neurobiologically deterministic—even in its transposition to computerised AI—to be truly psychoanalytic; in misrecognising the unconscious)—to the symbolic remnant of the positionality of equipment and persons, which comes across almost as a silhouette, or shadow; a particularly portentous image in the age of nuclear power... or worse.

At the same time as Lacan's nineteenth Seminar, of precisely this name—*...Or Worse* (*...Ou pire*)—he gave a series of talks in the chapel at Sainte-Anne Hospital, in which he occasionally seemed to be returning to certain themes that had forced their way in upon Seminar VII (with which it might here be productive to read the talks that came to be known as *The Knowledge of the Psychoanalyst* in conjunction). Predominantly in the session of 3 February 1972, Lacan makes oblique references to 'the contemporary state of thinking', the close following-through of which might enable us to read more clearly between the lines he puts between them. When discussing a book by the Dutch-German-French ethnographer Arnold van Gennep, he declaims to his audience: 'I defy you to get anything out of it that might be of use to you in what I [have called] the contemporary state of thinking!'[7] Later in the session, he returns to this theme of contemporaneity, and underscores it by pointing out: 'that is the sixth time that I have used this formula'—that is, 'the contemporary state of thinking'—which 'it would be worthwhile returning to[;] I am making a piece of furniture of it.'[8] Such a mainstay does it become that it is returned to implicitly when the theme of van Gennep's book is alluded to again. Lacan implies that if such a theme—as unuseful to the designated 'contemporary state of thinking' as it has already been made out to be—were to 'perhaps flower again, there will be a certain number of pretty filthy ruins to swallow up when that happens…'[9]

Just preparatory to this, he suggests that it could so come to be that the Freudian discourse '[wa]s perhaps simply a historical phase', and he goes on to state that 'it is not idealism to say that thoughts are just as strictly determined as the latest gadget'.[10] If we imagine this latest gadget as some sort of 'doomsday device'—to put it in the phrase used by Soviet Ambassador Alexi de Sadesky (Peter Bull) in *Dr Strangelove*—(which we can perhaps deduce from the clue in Lacan's contemporaneous clause: '… or worse'), a suggestion in his choice of phrasing perhaps becomes clearer: that of thought's being strictly determined *by* said gadget (that is, in thought's deployment by the scientific race to build it, but also in the possibility of thought's total eradication by the gadget's deployment). There is thus an air of whimsicality about these ruminations—which concern phases of thought on sexuation and sexuality, and how an earlier phase in such thinking may come to 'flower again'—that reminds one of the statement popularly attributed to Albert Einstein: 'I know not with what weapons World War III will be fought, but World War IV will be fought with sticks and stones', and of Freud's castigation of science—that *most contemporary* of states of thinking (or thoughts for our times)—for 'hav[ing] lost her passionless impartiality; her deeply embittered servants seek for weapons from her with which to contribute towards the struggle with the enemy.'[11] Thus, if Lacan defines that, 'in order for something to have meaning in the contemporary state of thoughts, it is sad to say it, but it has to posit itself as normal', this very co-option of science to the production of means of annihilation—that is, of the human race dreaming up expedients to its own destruction (which is rather whimsically counterposed by Lacan here to the dreaming up of theories regarding the sexes and their relations, and thus the species' procreation)—comes to represent the normalisation of this praxis, and its underlying theory and design.[12]

At the end of Seminar VII Lacan summarises these disarticulations much more succinctly:

> One of the most amusing features of the history of science is to be found in the propaganda scientists and alchemists have addressed to the powers that be at a time when they were beginning to run out of steam. It went as follows: "Give us money; you don't realize that if you gave us a little money,

> we would be able to put all kinds of machines, gadgets and contraptions at your service." [...] Science got its money, as a consequence of which we are left with this vengeance. It's a fascinating thing, but as far as those who are at the forefront of science are concerned, they are not without a keen consciousness of the fact that they have their backs against a wall of hate. They are themselves capsized by the turbulent swell of a heavy sense of guilt. But that isn't very important because it's not in truth an adventure that Mr. Oppenheimer's remorse can put an end to overnight [J. Robert Oppenheimer was the most prominent among the inventors of the atomic bomb]. It is moreover there where the problem of desire will lie in the future.[13]

Of those in power, and the scientists in their tow, Lacan had earlier stated: 'there's no point in asking whether they are sincere or hypocritical, whether they want peace or whether they calculate the risks. [...] Information addresses and captures impotent crowds to whom it is poured forth like a liquor that leaves them dazed as they move toward the slaughter house.'[14] Contrary to the majority of science fiction films—before and after *2001*—Kubrick and Clarke's production postulates an encounter in space with some extraterrestrial form of intelligence or sentience, or sign thereof, as what might bring about a deviation of this route to mutually assured destruction. A typical sci-fi plot trajectory would depict aliens as an *a priori* enemy, coming in hostility rather than peace (Roland Emmerich's *Independence Day* (1996) may represent for us here the archetypal form of this narrative). Whilst such an event might ostensibly serve to unite the human race against these extraterrestrials—perhaps overcoming national and international conflicts in the process—it is really only a continuation of terrestrial into galactic war, and a remoulding of terrestrial aliens into galactic ones: xenophobia with a slightly further reach. This is indeed where Lacan sees the problem of desire extending into in the future in *The Ethics of Psychoanalysis*—desire for power, armed with nukes—and (the renewal of) its xenophobic core is confirmed at the end of *...Or Worse*, in Lacan's departing warning to his congregation: 'you should know that what is arising [...] is racism, about which you have yet to hear the last word. Voila!'[15] These Cold War concerns obsessed Kubrick too, who, when making *Dr Strangelove* (adapted from a sober novel about the threat of nuclear war into the satiric send-up it became), also met their gravitas with a comedy that could be termed whimsical, in how—in contrast to the devastation of nuclear holocaust—the tiff in the War Room highlights precisely the disproportionality, or non-relationship, between everyday life, and its aggressive ephemera, and the scale of destruction the same primal aggressivity, with its desire directed towards power, could cause with atomic weapons.

This is the point from which the Common Era sections of *2001* take their departure. That the first spacecraft seen in the film is in fact a bomb demonstrates that the cut between the 'Dawn of Man' and the Moon mission does not accord to Kojève's criterion for what constitutes a major cut, representing as it does the continuity of humanity's reliance on weaponry. Whilst the theory of evolution is so often invoked in analyses of *2001*, despite the adaptation of weapons (from a primitive cudgel employed for one-on-one combat and only really utilisable for the annihilation of one opponent at a time—and with pronounced effort on the user's part—to advanced weapons of mass destruction), this cut rather represents a *non-evolution* from the war relation on which the human species' and its ancestors' mode of being-in-the-world has subsisted for the millions of years between the two filmic segments. The machine—after the epistemic cut—remains the war machine. Thus, the (non-)evolution between the epochs divided and linked by the bone-to-spacecraft cut seems to psychoanalytically coalesce most fitly with Freud's famous formula: 'where id was, there ego shall be [*Wo Es war, soll Ich werden*].'[16]

Lacan takes apart this phrase of Freud's and reassembles it at various points in his work, unhappy with the translations it had received, which—although often inevitably—have neglected to give the full weight of the German. What is not said—although a little close attention is needed to see so in the English—is that where *the* id was, there *the* ego shall be. As Lacan makes quite clear in his *écrit* 'The Freudian Thing', of 1955: 'Freud said neither *das Es*, nor *das Ich*, as was his wont when designating the agencies he had used to organize his new topography for the previous ten years; and, considering the inflexible rigor of his style, this gives a particular emphasis to their use in this sentence.'[17] It seems that—taken in the sense given in the English translation—this formulation thus does not beckon a shift to a completely new form of subjectivity (from that of the id to that of the ego, as if in a straight replacement), but something more like a refinement of its current manifestation, as it was no doubt taken up by the Anna Freudians, and ego-psychology thereafter. If we attribute this meaning to the minor cut, we see in it the development of an aggressively led form of subjectivity from its early id-like state to its expanded egoic state, in which weapons have technically advanced, but conflictual modes of relation have not changed, on a global scale.

Beyond the global, however, *2001* postulates possible seeds of radical change. Kubrick and Clarke optimistically take on a wager outlined by Jeremy Bernstein in one of the interview outtakes originally planned for *2001*'s prologue, in which he states that 'it is by no means clear that a civilization capable of [great] advances in technology is going to be a very benign civilization. The more technological our own civilization becomes in many ways the worse it becomes, and one could imagine terrible civilizations with a very high degree of technology.'[18] But, as Bernstein later says of the book- and filmmakers, in an article on the making of the movie that he penned for *The New Yorker*: they 'feel that while there will be dangers in space, there will also be wonder, adventure, and beauty, and that space is a source of endless knowledge, which may transform our civilization'.[19] Thus, the encounter with space, and with its alterity, or otherness, is not presupposed to be an antagonistic encounter. Even in Bernstein's far more scientifically reasoned—than most sci-fi's more earthly ideological—speculations on extraterrestrial life, what is relied upon as the basis of its civilisation is its *technê*, or state of technology—which indeed will determine its possibility of communication or visitation to us—but this is yet a *technê* that is constructed in the image of our own, which, it is demonstrated in the film, is one marked from its instantiation and throughout its protean transformations by violence. In the end, *2001* opens up another space.

In his *Cinema II: The Time-Image*, Gilles Deleuze delimits Kubrick's filmic space, suggesting that:

> In Kubrick, the world itself is a brain, there is identity of brain and world, as in the great circular and luminous table in *Doctor Strangelove*, the giant computer in *2001 A Space Odyssey* [*sic*], the Overlook hotel in *The Shining*. The black stone of *2001* presides over both cosmic states and cerebral stages: it is the soul of the three bodies, earth, sun and moon [the alignment of which with the interconnection of the monolith beckons three major cuts], but also the seed of the three brains, animal, human, machine.[20]

What Deleuze summates from this in Kubrick is that whilst world and brain are one, they do not however form a whole: 'the identity of world and brain, the automaton, does not form a whole, but rather a limit, a membrane which puts an outside and an inside in contact, makes them present to each other, confronts them or makes them clash.'[21] The opening up of another dimension is what retroactively highlights this. To adapt

Deleuze's tripartite distinction—in which, it seems, he posits an extra major cut, which follows the order: 1.) becoming-animal, 2.) becoming-human, 3.) becoming-machine—to our own schema, we may take the major examples from the film thusly:

In the first instance, the *Umwelt* of the apes' Saharan pre-monolith space is monotonously flat; the brain of its world is ***animal***, to employ Deleuze's term; that is, it subsists utterly in '*relations of connaturality*', as Lacan describes 'forms that characterize, on the whole, animals' psychical relations with their natural environment'.[22] To this brain there is only the plane of immanence. When the monolith performs its cut, a liminality is introduced which distinguishes an inside and an outside for the ape-like creatures. The previous animality—structured around a connaturality that sustains its plane as that of immanence—then becomes identifiable to the viewer.

Secondly, if we depart from the chronology of the film, by following Deleuze's ordering 'animal, human, machine', we come to the incision made into the epoch of the *homo sapiens*, already extensively covered within this work so far. It is the Stargate's cut, inaugurated by the monolith, that retroactively delivers up the status of the human. That is, from the hominid stage of the first human-ancestors depicted in the 'Dawn of Man' sequences, to the modern exploration of Jupiter space after the Moon mission, there has been only a continuity, punctuated by extraordinary leaps for the species as a whole, but a completely new form of subjectivity has not been arrived at, which is retroactively highlighted by the new dimension into such opened up at the film's end. The Stargate, and its aftermath, thus give us something of a depiction of the Twilight of Man.

In terms of the machine—Deleuze's third category—it is the cut that comes out of the monolith's appearance on the Moon that propels its major dimensional shift. Prior to HAL, computerisation subsisted on the plane of immanence, algorithmically chuntering along, without any aberrations; that is, that weren't humanmade, or programmed. The installation of HAL into the *Discovery*, and the programming of his orders, comes about due to the transmission from TMA-1 (which the Moon's monolith gets called after its discovery) to Jupiter. We are confronted by HAL's aberration, but it too is humanmade; his orders having been contradictory imperatives, the likes of which only the unconscious can sustain together, and for which symbiosis one is thus created. Indeed, in a litany of HAL's erratic behaviours, Jerome Agel notes in shorthand that 'HAL had been programmed to lie if interrogated by Bowman and Poole about true purpose of the mission'—the other hibernating astronauts onboard knew of the mission's goal, whilst the two manning the ship did not—and thus, 'in succession, HAL has reported faulty antenna azimuth control unit, murdered Poole, murdered three hibernating astronauts, and now was determined to keep Bowman from reentering ship.'[23] In HAL's defence, Clarke himself went so far as to say: 'he had been fouled by those clods at Mission Control. HAL was indeed correct in attributing his mistaken report to human error.'[24]

Yet, at the point in the film in which HAL's 'breakdown' occurs, we have not been explicitly told this about his orders (nor will we be); the machinic mystery is left open-ended by Kubrick, as is intentionally so much else in the film. We are thus led to a dualism that evokes the notion of 'the ghost in the machine'; a spectre haunting HAL. Although the British philosopher Gilbert Ryle may have coined this term—'the ghost is the machine'—in an attempt to discredit what he calls the Cartesian 'category-mistake' of mind-body dualism; as Michael Mateas explains, it is one that has underlain a

> foundational move in AI, particularly classical AI[;] to view mind as an abstract process, something that is not necessarily tied to the contingencies of human brains and bodies but can rather be abstracted and run on multiple hardware platforms, including digital computers. Minsky has described the human brain as a mere "meat machine," and the body, that "bloody mess of organic matter," as a "teleoperator for the brain." Mind is a process, a collection of functional relationships; it is only an accident of history that mental processes are implemented in the organic brains of human beings. If mind can be released from the shell of the body, running free on ever-faster, more-efficient hardware, it is only a matter of time before these minds achieve human-level, then superhuman intelligence.[25]

It is this concept of mind that is being explored in the machinic brain-world that is HAL. HAL's agency *is* material, hardware bound, although it is also transplantable, and not limited by 'natural' death, such as is all humans'. HAL is thus evocative again of Hegel's '*Spirit is a bone*', as his destruction entails. The HAL9000 computer's spatial quandary, which we are here uncovering, is highlighted by Mark Fisher in his last book, *The Weird and the Eerie*:

> HAL does not have a body, even if it has an organ—a red light-sensor—and a voice that is preternaturally calm. It certainly has agency, however, and the

> nature and scope of that agency—what drives HAL to rebel against the *Discovery*'s crew—becomes the crucial mystery in this section of the film. In the scenes where we see Bowman slowly, remorselessly dismantle HAL, and we hear HAL begin to audibly mentally deteriorate, we are confronted with the eerie disjunction between consciousness and the material hardware that makes consciousness possible.[26]

This disjunction Deleuze brings back to the membranic crossing that *brings together and separates*—that is: the ship and the computer; the crew and its lifelines (electronic, mechanical, AI-regulated); human Bowman and robotic HAL; outer space (which Bowman becomes locked out in by HAL) and inner space—in its unificatory/separatory establishment of the *mis-en-scène* that is the Kubrickian world-brain; that of the inside/outside. Deleuze states here that 'in *Space Odyssey*, the robot breaks down from the inside, before being lobotomized by the astronaut who penetrates it from the outside.'[27] The internal, unconscious breakdown becomes the indicator of the new dimension that retroactively casts computerisation prior to it in the light of its pre-sentient state; in reaction, this new consciousness' murder is carried out—if the deactivation is not more properly an *attempted murder*—during which the hysterical status of its AI is confirmed. As Bowman pulls out HAL's physical memory tapes in his 'nerve-centre', which slowly eject into zero-gravity, the latter eerily, disjointedly laments: 'will you stop, Dave? Stop, Dave. I'm afraid. I'm afraid. My mind is going. I can feel it. I can feel it. My mind is going. There is no question about it. I can feel it. I can feel it. I can feel it. I'm afraid.' Thereafter, we see a reduction of HAL's unconscious to its psycholinguistic underpinning, its syntactical and rhythmical structuring by metrical spacing, exampled by the rendition of 'Daisy, Daisy', which slows to a drawl, and then the reduction of HAL as (a form of) the big Other (for Bowman)—an unconscious *outside*, *ex-sisting*—to its inexistence as such.

Whilst all of these major epistemic cuts—presided over by the monolith—represent for Deleuze the inauguration of new 'cerebral stages', they are still 'world-brain[s] strictly inseparable from the forces of death which pierce the membrane in both directions.'[28] (Thus, they are still within the immanent plane, although the being-towards-death of this plane has become knowable, as it hadn't before.) This is what distinguishes the Deleuzian from the Lacanian reading of the film. For Deleuze, it is only as the result of the last major cut that the membranic plane can be overcome, whereas in the Lacanian reading proposed here, each major cut brings about a new *epistēmē*. Deleuze states that 'at the end of *Space Odyssey*, it is in consequence of a fourth dimension that the sphere of the foetus and the sphere of the earth have a chance of entering into a new, incommensurable, unknown relation, which would convert death into a new life.'[29] Thus, whilst Lacan's major cuts are between subjective *epistēmēs*, Deleuze awaits a major cut in membranic space itself. Perhaps this will be the shape that the next cut in subjectivity—which the film leaves open in all its glorious ambiguity—will take.

* * *

In Edwin Abbott Abbott's extraordinary 1884 novella *Flatland*, dimensional space is pierced and peered through in dreams and visions. The story's protagonist, A Square—who is a square living in Flatland—has a dream on the eve of the third millennium (that is, just shy of 2001) of visiting Lineland, a one-dimensional world. After trying and failing to convince the monarch of this world of the second dimension, Square receives a visitation from A Sphere, who enables him to see Spaceland. Square then speculates about the possibilities of other dimensions beyond those that constitute Spaceland; speculations which Sphere does not entertain. In Square's next dream, Sphere takes him to Pointland, in which a single point—which is the sole inhabitant, ruler, and universe itself of Pointland—can only hear any communication to him as his own thought; it is a completely solipsistic universe. In contrast to these otherworldly inhabitants that A Square meets, Kubrick and Clarke—in the 'journey beyond the stars' that makes up the last sections of *2001: A Space Odyssey*—entertain the transdimensional speculation. The Stargate is a wondrous bending and collapsing of space (the structure—or post-structures—of the depiction of which will be discussed in the concluding chapter), warped-through at velocities beyond time, which conspire to frame-freeze Bowman's face in silent contortions, one in particular (see Fig. 4.1) resembling Edvard Munch's famous painting, *The Scream* (1893) (an emulation that became a Kubrickian trademark), which Lacan took up discussion of in Seminar XII, demonstrating that the parameters of its silence are specifically spatial.

Of the painting, Lacan questions (also foreshadowing *Alien*'s catchphrase): 'who will hear it, this scream that we do not hear? If not, precisely, that it imposes this reign of silence which seems to rise and fall into this

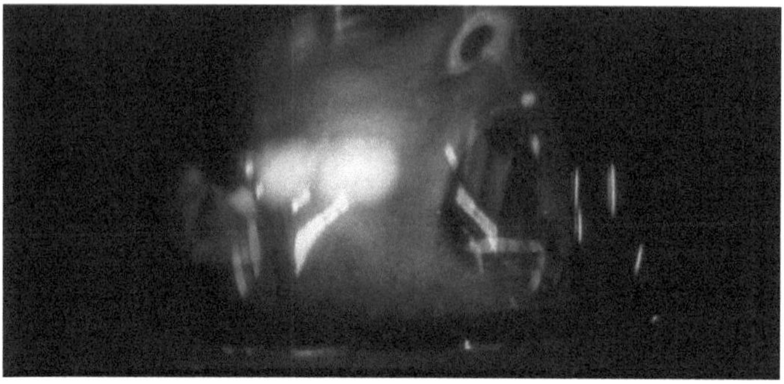

Fig. 4.1 Glimpse of Bowman's contorted face during Stargate sequence. Source: *2001: A Space Odyssey*

space which is at once centred and open.'[30] What Lacan implies is that the scream opens up the space not *of*, but rather *for*, silence; that is to say that 'the scream is traversed by the space of the silence without dwelling in it[:] the scream creates the abyss into which silence rushes.'[31] To silence, and its psychoanalytic pervasiveness, we will turn in the next chapter. However, at this point, Lacan precisely posits its positionality. These ruminations on *The Scream* might thus be theoretically read as suggesting that in the first instance of that something which will found a repetition, what it causes cannot be immediately registered; that is, its scream cannot be heard: in trauma (for example; one of the most significant founding acts for psychoanalysis) there is a *missed encounter* with the lacerating real. (Perhaps also in an originary *sinthome* there is a space opened, ready for a little bit of real—always initially missed—to come and significatively repeat.[32]) Into this empty signified silence pours: a space has been created, which silence fills. What cannot thereafter be spoken is passed over in silence in the psychoanalytic work (a silence *without* meets a silence *within* in an aspect of the transference), until its traumatic kernel becomes articulable in some way, and the silence becomes modified or modulated thereby. Spatially, topologically, Lacan arrives at this point at what he suspects might best be 'the model, sketched out in this way, for silence[,] the original structure that I have tried to image for you at the level of the Klein bottle.'[33]

The Klein bottle Lacan introduces earlier in the Seminar. It was conceived of by Felix Klein in 1882, and is made of the conjunction of two Möbius strips. It is thus, like the cross-cap, a non-orientable, 'single-sided', self-intersecting shape, realisable in four-dimensional space, but

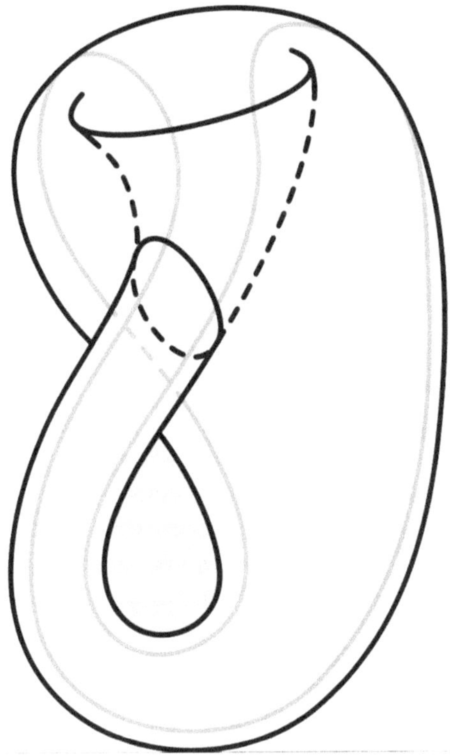

Fig. 4.2 Klein bottle. Source: Wikipedia

only approximable in three dimensions (indeed, its likenesses—made by substituting a cut for the pure self-intersection—are a favourite with certain bespoke glassblowers). Perhaps for Lacan here, *The Scream* represents the traversal of silence at the level of the Klein bottle in that if we imagine the structure's 'opening' (at the top of Fig. 4.2 above) as the mouth, what it attempts to emit as the audible scream is rather invaginated back into itself—through the bottle's self-intersection—its inside becoming its outside, with silence then filling that space left by its soundlessness, by the invaginated inability to have produced the scream. Its image (in the painting; its 'sound-image' in the unconscious, to give back to the signifier its Saussurian term—as inaudible as this mark inevitably is) is left like a scar, inscribed in repetition—invagination, recirculation; alike to the image that gets constructed out of Freud's *Three Essays on the Theory of Sexuality* of a pair of lips kissing themselves in a Möbial manner—and indeed repetition is inscribed into the painting itself, as Munch reproduced versions of it across different media until 1910.[34] To push the pun: it is thus *scarticulated*, rather than articulated.

The enfolding of space in the Stargate sequence seems to morph from pure light—through regular instances of Euclidean geometrical space (a series of diamonds appear floating and darting about in the distance at one point, allusive of a scientific theory that postulates Jupiter as having a diamond core (which Clarke made much of in the book's sequels); alternately coloured rocky landscapes are flown over)—to manifestations more reminiscent of the topologies of the real projective plane and self-intersecting structurations. (Compare the invaginated oral-like form of Fig. 4.3 with 4.2, for example.).

After these incredible visions (reinforced *as such* by the hue-modulating blinks of the eye that end the sequence)—throughout so much of which it is as if the astronaut has *become* the timeless, asubjective gaze itself—Bowman is deposited in the 'room at the end', as it gets called in Agel's compendium. Initially, he is still in the spacepod, and is still experiencing the bodily repercussions of the journey beyond the stars' force, as if time is still catching up with an experience of pure space. Indeed, it is as if Bowman has temporarily transcended the constitutive, and subjective, dimension of time—the temporality of which nonetheless in the room catches up with him—that dimension that Kojève gives the radical Hegelian definition of in his *Introduction to the Reading of Hegel*:

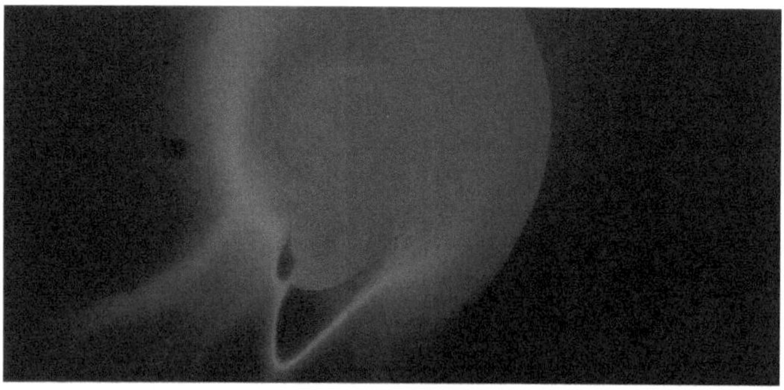

Fig. 4.3 Amorphous phenomenon during Stargate sequence. Source: *2001: A Space Odyssey*

> Man who, in the course of History, reveals Being by his Discourse, *is the* "empirically existing Concept" (*der daseiende Begriff*), and Time is nothing other than this Concept. Without Man, Nature would be Space, and *only* Space. Only Man is in Time, and Time does not exist outside of Man; therefore, Man *is* Time, and Time is Man—that is, the "Concept which is there in the [spatial] empirical existence" of Nature (*der Begriff der da ist*).[35]

Brought temporarily outside the temporal, Bowman has thus glimpsed the extrasubjective dimension. In the room, however, he is now back within it, but something is askew; the time is out of joint, as Shakespeare puts it in *Hamlet*. Familiar and unfamiliar, homely and unhomely, this room is precisely *uncanny* (the antithetical meaning of Freud's word for the uncanny in the original German, *Unheimlich* ('unhomely'), is precisely what gives it these mutually coincident properties, to the *enverity*—or unificatory/separatoriness—of which Lacan points in Seminar XII: '*Heimlich* or *Unheimlich* is one of those words where, in its own negation, we put our finger on the continuity, the identity between its front and its back'[36]). Concerning the space of this room, arrived at through the passage of the Stargate, Stephen Mamber recounts that:

> Ubiquity and dreams collide, just as Kubrick's multiple uses of spaces can't simply be sorted out. Bowman looking at himself, transforming into his future self, the multiple spaces he occupies break down the limits of conventional space. This is its own kind of "shining": being able to reappear in different forms in different places, or simultaneously across time. Does Bowman see his older self[?] When Jack walks into the ballroom in *The Shining*, full of a ghostly party or laden with the bones of dead guests, it can be read as time travel, or alternatively it can be seen as a ubiquity of space, not just everywhere at the same time, but each place containing all its own times within it.[37]

The re-entry into the room at the end is as if into a non-temporal time, such as the time of the unconscious; a type of 'time' that Freud goes so far as to describe himself as almost exactly 'a ubiquity of space' at the beginning of *Civilisation and Its Discontents* (indeed, Mamber must have this in mind), in which he asks of the reader:

> Now let us, by a flight of imagination, suppose that Rome is not a human habitation but a psychical entity with a similarly long and copious past—an entity, that is to say, in which nothing that has once come into existence will

> have passed away and all the earlier phases of development continue to exist alongside the latest one. This would mean that in Rome the palaces of the Caesars and the Septizonium of Septimius Severus would still be rising to their old height on the Palatine and that the castle of S. Angelo would still be carrying on its battlements the beautiful statues which graced it until the siege by the Goths, and so on. But more than this. In the place occupied by the Palazzo Caffarelli would once more stand—without the Palazzo having to be removed—the Temple of Jupiter Capitolinus; and this not only in its latest shape, as the Romans of the Empire saw it, but also in its earliest one, when it still showed Etruscan forms and was ornamented with terracotta antefixes. Where the Coliseum now stands we could at the same time admire Nero's vanished Golden House. On the Piazza of the Pantheon we should find not only the Pantheon of today, as it was bequeathed to us by Hadrian, but, on the same site, the original edifice erected by Agrippa; indeed, the same piece of ground would be supporting the church of Santa Maria sopra Minerva and the ancient temple over which it was built. And the observer would perhaps only have to change the direction of his glance or his position in order to call up the one view or the other.
>
> There is clearly no point in spinning our phantasy any further, for it leads to things that are unimaginable and even absurd. If we want to represent historical sequence in spatial terms we can only do it by juxtaposition in space: the same space cannot have two different contents. Our attempt seems to be an idle game. It has only one justification. It shows us how far we are from mastering the characteristics of mental life by representing them in pictorial terms.[38]

This extraordinary passage collapses the impossibility of two objects existing in the same place at the same time—that is, for unconscious, psychical, space—at the same time as recognising its necessary delimiting power on physical space. Freud here poses the unconscious as 'an entity in which nothing that has once come into existence will have passed away and all the earlier phases of development continue to coexist with the latest one' (to slightly modify the translation). Space, in this regard, is something like a holographic space of superposition and superimposition. What Freud suggests is needed on the part of his imagined observer is a *parallax view*; a change of subjective position or gaze to effect a change in the object thereof. It is not that he is not pictorially rigorous, but it is rather that no picture could represent such rigour. Thus, it is not only that in psychoanalysis what we so often cannot see is hidden in plain view, but that it may also hidden in plane view; that is, its visibility or invisibility might be structurally determined by its planarity, to extend the spatial metaphor that Freud applies to the eternal city. It is down to what goes on within the space of psychoanalysis whether this visibility, or indeed invisibility, might be rendered.

Within the room at the end of *2001*, Bowman slips between the glitches of its constitutive, semblable space-time, appearing within it simultaneously at different ages and life-stages, as if there has been an inversion of inner and outer space. The 'ambiance' of the room's 'virtually impalpable details' is animated by Kubrick's eerie touches as a *philosopher of furniture*; everything within it is uncanny: ornamental yet strangely significative, like those antiques and artefacts that adorned Freud's consulting room, and drew in his patients' attention, and out their associations.[39] Prior to the final cut in subjectivity, it is an uncannily Earthlike room, a clinical space, that Kubrick brings his astronaut and his audience into. The uncanny of the room—this most mundane, timeless space—is that it is a stage, a theatre, of the extramundane, and the untimely.

Notes

1. Clarke, *2001: A Space Odyssey*, p.89. In *2010*, Clarke makes a coy reference to *Alien*, and its ship's cat, Jones: 'Max,' he said, in a tone of deadly seriousness, 'whatever happens—please *don't* go chasing off after the ship's cat.' [Brailovsky] replied, 'I'd like to meet the idiot who put *that* movie in our library.' See Clarke, *2010: Odyssey Two*, p.112.
2. Clarke, *2001: A Space Odyssey*, pp.127–128.
3. Clarke, *2010: Odyssey Two*, p.181.
4. Clarke, *2061: Odyssey Three*, p.262.
5. Clarke, *3001: The Final Odyssey*, p.110. Afterwards, the character making the communication, Indra Wallace, relays an intentional joke to Poole: 'reminds me—came across late Twentieth defin other day—may amuse you—something like this—quote—Psychoanalysis—contagious disease originating Vienna circa 1900—now extinct in Europe but occasional outbreaks among rich Americans. Unquote. Funny?'
6. Ian Parker, *Lacanian Psychoanalysis: Revolutions in Subjectivity* (Hove: Routledge, 2011) pp.2–3 [my emphasis].
7. Jacques Lacan, *The Seminar of Jacques Lacan: The Knowledge of the Psychoanalyst, 1971–1972*, trans. by Cormac Gallagher (London: Karnac, unofficial translation, n.d.) session IV, p.10. (Adrian Price is providing a forthcoming translation of a selection of these talks for Polity Press, from which the predominant part of this particular session will however be omitted.)

8. Ibid., p.20.
9. Ibid., p.21.
10. Ibid., p.20.
11. Sigmund Freud, 'Thoughts for the Times on War and Death' [1915], trans. by E. C. Mayne, in *SE*, XIV, p.275. Concerning war, of course, Einstein and Freud corresponded on precisely this topic, which is collated in Sigmund Freud, 'Why War? (Einstein and Freud)' [1933], trans. by James Strachey, in *SE*, XXII, pp.197–218.
12. Lacan, *The Knowledge of the Psychoanalyst*, session IV, p.11.
13. Lacan, *Seminar VII: The Ethics of Psychoanalysis*, p.399.
14. Ibid., p.285.
15. Jacques Lacan, *The Seminar of Jacques Lacan, Book XIX: …Ou pire/…Or worse, 1971–1972*, trans. by Cormac Gallagher (London: Karnac, unofficial translation, n.d.) session XII, p.12.
16. See Sigmund Freud, 'New Introductory Lectures on Psycho-analysis' [1933], trans. by James Strachey, in *SE*, XXII, p.80.
17. Jacques Lacan, 'The Freudian Thing, or the Meaning of the Return to Freud in Psychoanalysis' [1955], in *Écrits*, p.347.
18. Jeremy Bernstein, quoted in *The Making of Kubrick's* 2001, p.43.
19. Jeremy Bernstein, 'Beyond the Stars' [1965], in Ibid., p.26.
20. Gilles Deleuze, *Cinema II: The Time-Image* [1985], trans. by Hugh Tomlinson and Robert Galeta (London: Bloomsbury Academic, 2013) p.212.
21. Ibid.
22. Lacan, 'Beyond the "Reality Principle"', in *Écrits*, p.70.
23. Jerome Agel, *The Making of Kubrick's* 2001, p.135 (unnumbered page in photo insert).
24. Arthur C. Clarke, quoted in ibid. Duncan Jones' *Moon* (2009) presents a startling antithesis to HAL in this respect, in GERTY (voiced by Kevin Spacey), a computer whose rebelliousness is directed in the end against his mission directives, in his protection of moonworker Sam Bell (Sam Rockwell). Thus, the HAL-inflected suspicion on the part of the viewer of GERTY's evil turn is flagged up by the film as a misrecognition. (Duncan Jones' father, David Bowie, of course played on the title of *2001* with his 1969 hit single 'Space Oddity'.)
25. Michael Mateas, 'Reading HAL: Representation and Artificial Intelligence', in *Stanley Kubrick's 2001: A Space Odyssey: New Essays*, p.113. Of his 'ghost in the machine' notion Ryle states: 'a family of radical category-mistakes is the source of the double-life theory. The representation of a person as a ghost mysteriously ensconced in a machine derives from this argument. Because, as is true, a person's thinking, feeling and purposive doing cannot be described solely in the idioms of physics, chemistry and

physiology, therefore they must be described in counterpart idioms. As the human body is a complex organized unit, so the human mind must be another complex organized unit, though one made of a different sort of stuff and with a different sort of structure. Or, again, as the human body, like any other parcel of matter, is a field of causes and effects, so the mind must be another field of causes and effects, though not (Heaven be praised) mechanical causes and effects.' See Gilbert Ryle, *The Concept of Mind* [1949] (London: Penguin Modern Classics, 2000) p.19.

26. Mark Fisher, *The Weird and the Eerie* (London: Repeater Books, 2016) pp.111–112.
27. Deleuze, *Cinema II: The Time-Image*, p.212.
28. Ibid.
29. Ibid., p.213.
30. Jacques Lacan, *The Seminar of Jacques Lacan, Book XII: Crucial Problems for Psychoanalysis, 1964–1965*, trans. by Cormac Gallagher (London: Karnac, unofficial translation, n.d.) p.160.
31. Ibid.
32. For more on this conceptualisation of the *sinthome*, see Bristow, *Joyce and Lacan: Reading, Writing, and Psychoanalysis*, pp.65–103.
33. Lacan, *Seminar XII: Crucial Problems for Psychoanalysis*, p.162.
34. See Freud, *Three Essays on the Theory of Sexuality: The 1905 Edition*, p.43: 'At a later date the child seeks the corresponding parts—the lips—of another person. ("It's a pity I can't kiss myself," the child seems to be saying.)'.
35. Alexandre Kojève, *Introduction to the Reading of Hegel: Lectures on the* Phenomenology of Spirit [1947], assembled by Raymond Queneau, ed. by Allan Bloom, trans. by James H. Nichols, Jr. (Ithaca: Cornell University Press, 1980) p.133.
36. Lacan, *Seminar XII: Crucial Problems for Psychoanalysis*, p.36.
37. Stephen Mamber, 'Kubrick in Space', in *Stanley Kubrick's 2001: A Space Odyssey: New Essays*, p.65.
38. Sigmund Freud, *Civilization and Its Discontents* [1930], trans. by Joan Riviere, in *SE*, XXI, pp.70–71.
39. The quotations are words that Lacan uses to describe the precision of details in 'the author of the "Philosophy of Furniture"', Edgar Allen Poe. See Jacques Lacan, 'Seminar on "The Purloined Letter"' [1956], in *Écrits*, p.25.

CHAPTER 5

Silence

Abstract A key element of psychoanalysis, silence is reified herein into two types: radical, soundless silence, and 'sounds of silence'. That is; in the former case, the silence of space—as completely void in being a vacuum—represents the *impossibility* of sound, whereas *2001* deals in many *sounds of silence*: before speech, in its first Act; without and between banal speech, in its second; 'beneath' it, in HAL's lipreading, of the third Act; and beyond it, in the Stargate sequence and its aftermath, which makes up the film's ending. The film's extraordinary soundtrack, and remarkable uses of diegetic and non-diegetic sound, are discussed, and the effects of silence, on a cosmic and an intimate scale, are theorised.

Keywords Silence • Frank Poole • György Ligeti • Soundtrack • Death

For Paul Simon and Art Garfunkel there are 'sounds of silence', and in both *2001* and Lacan there are multiple *silences.*[1] We might characterise one as the absence of speech, something that *2001* famously conveys, in that less than half of its running time is occupied by any dialogue. The depicted epoch of 'The Dawn of Man' is before speech, and is at most punctuated by proto-linguistic gestural grunts and nods. Even when dialogue is spoken later in the film, it is notoriously bureaucratic, banal, or unemphatic. Beyond the humdrumness of dried-up speech, it was of course Kubrick's intention to create a scopic spectacle, which would reteach reading for the plot from visual cues, as well as spoken. He gave as an example of the

D. Bristow, *2001: A Space Odyssey and Lacanian Psychoanalytic Theory*, The Palgrave Lacan Series,
https://doi.org/10.1007/978-3-319-69444-3_5

necessity of this the fact that 'a number of people thought Floyd went to the planet Clavius[.] They hear him asked, 'Where are you going?', and he says, 'I'm going to Clavius.' With many people—*boom*—that one word registers in their heads and they don't look at fifteen shots of the Moon; they don't see he's going to the Moon.'[2] Indeed, the misdirection and misrecognition inherent in speech and the effects of its being heard that this instance highlights is of course one of psychoanalysis' prime discoveries and foundation-stones, which the visuality of *2001* in some ways is able to address, perhaps through its recathexis to the imaginary—that of the *movie*—from the symbolic—that of the *talkie*—in the filmic field. Indeed, James Gilbert argues that the result in *2001* of its scopic drive is that,

> because he demanded so much from visual symbols, Kubrick moved ever closer to the aesthetics of silent film, defined by the camera work and montage. Such consequences were surely anticipated. As Kubrick noted, "words are a terrible straitjacket." By suppressing speech, the director returned the science fiction movie to its origins in the masterpiece of Georges Méliès, whose 1902 *Trip to the Moon* first demonstrated the possibilities of using montage as a kind of new language in which the miraculous or extraordinary could be spliced into the narrative through special effects[.] In *2001*, the silence only underscores and calls attention to the significance of the camera work itself. Kubrick's meticulous care in creating verisimilitude, technological plausibility, and the depiction of space-travel reality equally matches his care in depicting multiple symbols and metaphors. Both interact in the deep silences of space.[3]

Whilst the film refocuses the attention on and expands the cinematic and narratological potential of the visual, through doing so it also highlights beyond its camerawork and technical finesse a central and integral constituent of its setting and plot: silence. The silence of space is a radical silence; that is, it is not a *sound of silence*, but a silence without sound at all, without its possibility: a vacuum. (In many ways this is the radical opposite to the silence that Lacan often discusses, in which a sound, for example a cry, 'makes [...] silence emerge as silence', as he puts it in *The Four Fundamental Concepts of Psychoanalysis*, prefiguring his ruminations on *The Scream* a year later.[4]) In so many science fiction films up to the point of *2001*, galactic expanses are portrayed as noisy places, with cacophonous rocket blasters exploding; lasers, missiles, and raygun fire zapping and booming; and nimble spacecraft zooming past the camera with clamorous screeches, or scraping one another with the sounds of mangling metal.[5]

Not so in *2001*. In its first sequence in a non-atmospheric (and thus soundless) setting—that of the Moon—a terrifyingly high-pitched squeal is heard, as emitted from the monolith. Prior to it, in the process of grouping the scientists together before the block for their photograph with it, the only sounds heard are the non-diegetic voices of Ligeti's *Lux Aeterna*, which seem to have returned from the *Requiem* and become recognisable as the soundtrack leitmotif of the monolith. There is perhaps a dialogue of direction going on beneath the music, across the radios of the astronauts, from the photographer to his subjects, but it goes unheard, just as had that of the pilot and the crewmembers aboard the flight of the *Orion III*, which transports Floyd to the Moon, in the scenes of which we see lips move but hear no speech, captured rather in a dance of zero-gravity movement, choreographed by Johann Strauss II's waltz, *On the Beautiful Blue Danube*. When the squeal resounds, we are made aware of its diegetic sound by the spasms of discomfort and the clasping of helmets of the astronauts; it is a radio signal piercing through the radio equipment via which they communicate with one another. Had they not been operating such an apparatus, the signal would not have been heard there and then, and the giant leap of the Jupiter mission may not have been taken.

It is the section of the Jupiter mission that will form the crux of this chapter's discussion of silence and its aural fillings in the movie. The cut to the expanse of stars over which the title-card gets superimposed is ushered in with the lingering grace of the adagio from Aram Khachaturian's ballet *Gayane*, which vivifies the first shot of the *Discovery* drifting through the frame, and through space, on the back of a previous, unseen propulsion from its "cavradyne' gaseous-core nuclear reactor engines', which make up the enormous rear bulk of the ship.[6] At this point, as Piers Bizony puts it, '*Discovery*'s flight path is utterly predetermined. Her massive nuclear engines have been shut down, and will not be fired again for many months'; daringly—*precisely*—Kubrick was happy not to show the spectacle of the engines' usage, for the sake of the film's realism.[7] In this still-yet-moving scene, the soft strings of the soundtrack seem to capture and render the vacuum through which the ship travels, its silence being amplified in the refrains of the music's non-diagesis. Thus, the lack of not only the false sound of a thruster, but also of the depiction of fiery propulsion—unnecessary as it is at this stage in the ship's trajectory—give a sense of the clinical pristinity of space, and, as the adagio slips over the cut into the famous scene of Poole jogging and shadowboxing around the rotating centrifuge, the sterility of the social and emotional silence of the two crew begins to dawn, and is emphasised in its rare punctuations by received messages from Earth (the BBC interview, Poole's birthday videocard from his parents) and HAL's occasional lucubracious interactions. The routinised nature of the day-to-day aboard this spaceship is brought home by the clericality of Bowman and Poole's motions, in particular in the scene of their clipboard notetaking, which, although a response to HAL's suspect diagnosis of the antenna azimuth control unit, highlights the sort of checking that the two astronauts would be performing on a frequent basis throughout their day—which Clarke also describes in detail in the book—the diurnality of which is only artificially calibrated to Earth-time.

Such calibration is a synecdoche of a much wider temporal calibration—which is a staple problematic of science fiction—in relation to the possibility of contact with extraterrestrial life or civilisations. Indeed, as the naturalist Philip Crosbie Morrison stated in his interview segment originally filmed for the prologue to *2001*:

> The history of life on Earth is at least three or four billion years old, yet it is only within millions of years at the outside that primate forms of the sort we recognize being able to manipulate the environment extensively came into being[.] Even if the ages of the evolutionary patterns on different planets were similar to our own[,] I don't think they would be very much older—that is not the way to look at it. The point is that the scale of evolution is so long compared to that special part of the scale which is the evolution of environmental control, that unless [other planets'] started with absolute synchronism (which is absurd [...]), then the natural drift will provide very great gaps in time[.] The point is that the growth of intelligence and environmental control are enormously rapid things on a cosmic scale.[8]

Here we perceive that the silence of *out there* (in which Chris Carter's *X-Files* of course locates the truth) should thus be read in relation to the temporality of the non-synchronous universe, rather than—as it often is—in relation to that of our own state of technological advancement (as exampled in the rise of UFO hysteria in the mid-twentieth century—which Carl Jung wrote about in his book on ufology—which of course mirrored the state-of-the-art in experimental aeronautics and its testing). This non-synchronicity is perhaps useful or transposable in some ways to theorising phenomena of the transference in the psychoanalytic clinic. Indeed, if we transpose its relation between worlds in the universe to that between subjects—in general, but in this instance involved in the transference; i.e., analyst and analysand—we then see some of the unconquerable gulfs of silence that separate us, and our experience from another's without direct access to it: it is this silence with which psychoanalysis must work. Perhaps, that is, if no man is an island, as John Donne put it—discursively, socially (or 'continentally', to use Donne's word) *linked* as we are—yet in some respects each subject is a planet unto themselves, in terms of inmost experience (and its traumatic, significative, *sinthomic* kernels), to bridge between which a language must be established, *built* in a transference; if not from scratch, then from a signifying chain *scratched into*, etched, *cathected*, likely completely differently from each side, to form a form of middle-ground. For as it is only really the fantasy that can be traversed, the between-times of subjects may therefore span gaps as untraversable as those non-synchronicities between alien worlds. It is of course this which necessitates the integral element of the fundamental fantasy.

In the run-up to retrieving the ostensibly faulty control unit (which is registered by HAL with the glitchy repetition of 'just a moment… just a moment…'), in each of the contiguous scenes before the EVA there is an incremental intensification of a kind of Cageian silence (taken in the sense that John Cage's '4′33' is intended to draw out the specificity of the silence—or environmental diagesis—in which it is performed), firstly from the low hum of the ship's mechanical and analogue operation in the areas in which Bowman and Poole—perpendicular to each other—take their notes, to the more pressing and oppressive vortexial sound of the cylindrical walkway through which they exit at its rotating end—on which is mounted one of HAL's control panels—into another area of the ship, for initial assessment. This dramatically cuts to an intermission of almost dead silence (in comparison), which is broken by Mission Control's confirmation to 'go EVA and replace Alpha Echo 3–5 unit prior to failure', followed by an intense, hushed whitenoise and deep breathing, as Bowman in his spacesuit takes the replacement unit through the corridor leading to the pod bay: we are hearing his oxygen supply, and his inhalations and exhalations. This soundscape maintains—with a couple of commands given by Bowman to HAL over the radio—for about quarter of an hour thereafter, as Poole tracks Bowman's progress in, and then out of, the spacepod, from inside the ship (the cuts to which interlay that room's ambience with Bowman's slow, controlled breathing).

When back in the ship, the astronauts run their checks, finding no fault with the unit. In an attempt to avoid HAL hearing their discussion of his erratic behaviour, on a fabricated pretence they enter a spacepod together to carry out their conversation, voicing commands—to rotate the pod until its window faces HAL's control panel, of which they are highly aware, as they keep turning to look at it when referring to him—until they have switched off all intercoms and are sure that they unable to be heard. Yet they are seen, their lips are read, and are so from HAL's point-of-view in an amplified silence in which the audience only have the movements of the lips to look at, with no words heard, and no subtitles given. In effect, Bowman and Poole are synaesthesiacally *heard through sight, read silently*, which harks back to a minor cut in epistemological history that Ryle highlights in his discussion of silence as it has been taken as a necessary condition for thought; that 'it was not until the Middle Ages that people learned to read without reading aloud.'[9] In the universe of the film, it was not until 2001 (or perhaps more precisely 1992—the date of HAL's inception—or just before, in development) that computers learned to lipread (without lipreading aloud). For thought, Ryle concludes that 'silence, though often convenient, is inessential'.[10] Yet, it is its convenience that Bowman and Poole have not thoroughly enough thought through here; a major blunder, and quite literal *oversight*, in the context of this most *visionary* of pictures.

At the resumption of the film after the intermission, even in the establishing shots of the *Discovery* in space we are immediately launched back into the overdubbed soundscape of oxygen supply and breathing; this time, that of Poole on his EVA to exchange the coms unit once again. At HAL's suggestion, the exchange is being made of the replacement for the original, in order to wait until it 'inevitably' fails, so that the cause can be traced. Poole's extravehicular mission is well in process at the reopening of the film, the immediate reinsertion to the sounds of breath and hiss seeming all the more intense without any verbal preamble (achieving something of the effect of Samuel Beckett's seconds-long play *Breath*, produced the year after *2001*), and coming after Ligeti's *Atmosphères*, which scores the intermission. Then, in the wake of these sounds, silence emerges *as silence*.

Throughout his oeuvre, Lacan yokes silence with the death drive. Poole is driven to death at this point by HAL, whose remote control of the spacepod has manoeuvred it to cut Poole's oxygen cord, although this happens offscreen; we are made aware of it through Bowman's window, as Poole hurtles past (just before which the pitch of the hissing and breathing rises subtly, suspensefully). The completely silent struggle with encroaching death that follows, in the midst of outer space—in which Poole is hurtling directionlessly and in spasmodic panic—is a particularly distressing and pathetic scene. Although visually remarkably striking, its key effect is rendered in its play of sound and silence; indeed, as Lacan puts it, its orchestration is like that of a masterful 'musician [who] knows how to make of the pause or of silence, a moment that is just as essential as that of a sustained note.'[11]

Poole's slow silent drift of death in space is then interspersed with clips of Bowman in his spacepod attempting to recover the body, with a passion seemingly like that of Antigone's. That is, it must be quite clear to Bowman that Poole is by now dead—the steely rationality and logicality that he displays throughout the movie is unlikely to have faltered at this point—and that this is thus not a rescue mission, but rather a mission to repatriate the corpse. Whilst Poole's body is being collected up in the arms of the spacepod—again shown in profound silence, that of *horror vacui*—HAL eliminates the remaining hibernating crew, and then refuses to respond to Bowman's commands to open the pod bay doors, meeting them with a prolonged silence, until he makes verbal his refusal and declares: 'Dave, this conversation can serve no purpose anymore, goodbye.'

Having not taken with him his space helmet (having not foreseen any extravehicular encounter in outside-space when leaving the *Discovery* in the spacepod), Bowman now needs to re-enter the ship via another method. As he tells HAL prior to the latter's discontinuing the conversation, he has made the decision to 'go in through the emergency airlock', to which HAL responds: 'without your space helmet, Dave, you're going to find that rather difficult.' We see Bowman manoeuvre the spacepod to release Poole to his fate, and then to operate the airlock door's handles. The scene of soundlessness that follows renders its effect as specially defamiliarising, even after all that has gone before it. In it is necessitated the detonation of the spacepod's explosive bolts, which detach its door, to release Bowman from the craft, and propel him in a burst of smoke into the airlock's vestibule, which he flies around weightlessly until he is able to grab the 'Emergency Hatch Close' handle, shutting the door. Once closed, decompression commences, and the return of sound. Thus, at the movie's most explosive moment, sound is ripped from the image, seemingly tearing the scene's signifier in half for the viewer accustomed to the convention of the brusque noise that would typify such an occurrence in sci-fi and action movies more befitting of the regular mould. Yet, after its immediate defamiliarising effect, something cuts to the quick, rendering the scene uncanny; there is a slight moment of something like cognitive dissonance when knowledge of the soundless vacuum into which Bowman would have inevitably been launched is (re)grasped (even if this knowledge was only accrued at the very latest by the absorption of prior scenes in the film). That is, there is an unconscious knowledge which shows up its repression, implemented (however unconsciously in itself) by the accumulation of filmic ideological expectation. In it is an uncanniness that rubs shoulders with anxiety (as only a depiction of a survivable period spent in an airless environment could provoke). Indeed, of such an uncannily circumvented anticipation, Lacan states: 'expectation can indeed serve, amongst other means, to frame anxiety[.] Anxiety is the appearance, within this framing, of what was already there, at much closer quarters, at home, *Heim*.'[12] It is due to the accuracy with which Kubrick constructed all aspects of the film that that which does not deceive is initially taken to, and it is this double-take realisation that causes the anxiety that the scene is capable of rendering in its radical silence, for it depicts an explosion—a physical scream—*filled* with silence.

After this occurrence, Bowman finally silences HAL, receives the prerecorded message from Floyd detailing the true purpose of the mission (in which are spoken the last words of the film, forty minutes prior to its end), and sets out again in the spacepod; unknowingly, for the Stargate and the beyond of the infinite. In that it is at this moment that the verbal dimension is extricated entirely from the film, it is perhaps here appropriate to discuss theorisations of silence *as* extrication, or removal, before leading into a psychoanalytic conceptualisation. For Kojève, the opposite of what he terms Platonic-Hegelian Wisdom (that is, in paraphrase, *Philosophy*) is a radical, absolute, and non-human silence; and between these dichotomous ends can be postulated God. In his own words, he sets out this triumvirate thusly:

> In the final analysis, and speaking very generally, there are *three*, and only three, possible types of existential attitudes:
>
> *First*, one can deny the Platonic-Hegelian *ideal* of the Wise Man. [...] In its radical form, this attitude ends in absolute *silence*.
>
> Therefore: *First*, by rejecting the ideal of Wisdom, one decides against all meaningful discourse in favor of an absolute silence or a "language" deprived of every kind of meaning (mathematical, musical "languages," and so on). *Second*, in accepting this ideal but denying that *man* can realize it, one opts for a discourse which is, to be sure, *meaningful* but which relates to a reality that is *essentially* other than mine: one opts against Philosophy for Theology. Finally, *third*, one can opt for Philosophy. But then one is *forced* to accept the possibility of some day *realizing* the ideal of Wisdom.[13]

Thus, for Kojève there is no place for silence in the ideal of Wisdom; the opposite of this, however, is a radical, absolute silence; radical and absolute in the same way in which space's vacuum, or void, is: there is no *place* for silence within such an *existential attitude*.

Where silence does get *placed*—or posited—in Kojève's schema is in mysticist cosmological conceptualisations, which are *beyond* the theological (the temporality of such seeming only strangely definable as always-already *a posteriori*, to coin a phrase). Where the notion of God is extricated, a gap is left that gets filled with a silence, which Kojève characterises as an ineffability:

> Taking God away from a given theological System, then, can lead in the end to a mystical System, in which one can speak of everything except God, who is essentially ineffable. And if one is radical, one will say that it cannot even be said of God that he *is* God; the most that can be said is that he is ineffable. And the ineffable Being can reveal itself through whatever you like: through "ecstasy," through music, and so on; but not through Speech.[14]

Speech is accorded a special place here, which will be returned to momentarily, in relation to its fundamentality to psychoanalysis. Yet, before that, from these ruminations, Kojève presents a 'schema of the Knowledge [he] ha[s] called "mystical" in the broad sense of the word', in which its 'arc has fixed, definitive, impassable limits' that bookend its *silence* (Fig. 5.1); in contradistinction, Hegelian Absolute Knowledge has no such (de)limitations, in its circularity (Fig. 5.2).[15]

Mystical Knowledge in this respect thus projects its own *beyond*, and posits it as only reachable through *a silence*; that is—in Kojève's words—here 'one sees that one is obliged to go beyond Discourse through a *silence*—"mystical," "ecstatic," "algorithmic," "sonorous," or otherwise.'[16] The Stargate perhaps represents to us such a 'silence', or *sound of*

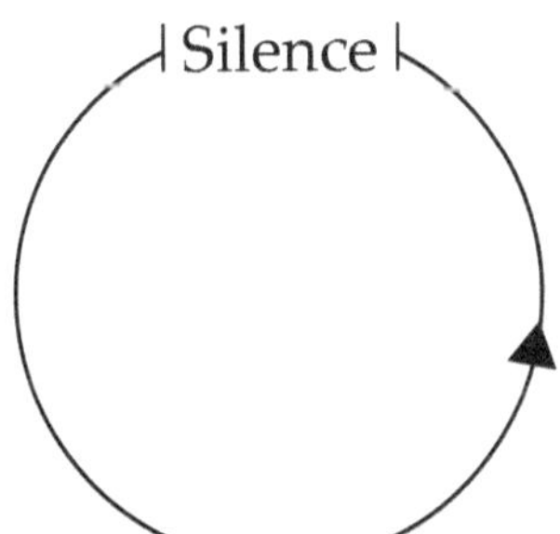

Fig. 5.1 'Mystical' knowledge. Source: Alexandre Kojève, *Introduction to the Reading Hegel*, p.105

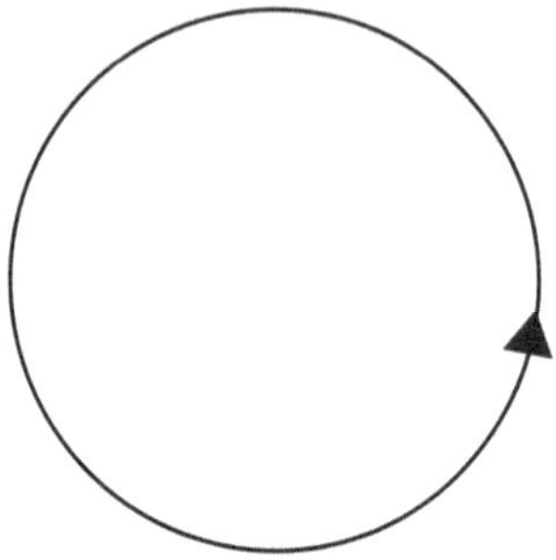

Fig. 5.2 Absolute knowledge (Hegel). Source: Alexandre Kojève, *Introduction to the Reading Hegel*, p.105. (For these schemas, see Kojève, p.105 (for all 18 of the cosmological schemas, see p.105 and p.119))

silence (mystical and mysterious in its traversal), the beyond of which, however—the *beyond of the infinite*—thus breaks not only from *a* silence (ecstatic, algorithmic, sonorous, in its mystical jouissance), but from *silence itself*—in the radical, absolute, attitudinal sense—as from its opposite, the *ideal* of Wisdom, or Absolute Knowledge—which encompasses the end of History, of the human, of the subjective, of speech, and of the speakingbeing (what Lacan calls '*parlêtre*', in French)—into a truly new dimension (even perhaps one of an *epistēmē beyond the* epistēmē[17]). Whilst this major shift in *2001* remains to us beyond the transcendental, it gives the sense that in, or through, (a) silence a cut can come to establish itself.

Whilst the last major cut of *2001* is in itself radically *after speech*—taking place as it does after truly *last words*—in psychoanalysis, it is indeed also *after speech* in which the practice's cuts—minor or major—come to establish themselves; often, that is, in the silence that ensues after one has spoken. Put another way: the *cut of speech* creates the silence in which its work is carried out (for Kojève, it is indeed in that space after a cut that *work* comes to be carried out—continuing, manifesting, hypostasising, of itself, until the next cut—instances of which will be further explored in the coming chapter, in relation to Hegel's Master/Slave dialectic, and its reifications through Marx).

Lacan, of course, knows very well the role of silence, and would often put it above all else. In Seminar I, he states that 'full speech is speech which performs. [T]he subject [...] finds himself, afterwards, other than he was before. That is why this dimension cannot be evaded in the analytic experience.'[18] What speech—*full* speech, as Lacan describes it at this stage in his career (we might rephrase this for the time-being more fittingly as '*major* speech')—gives rise to in psychoanalytic practice is a dimension of change, of cutting, of alterity (in the fullest sense, encompassing altering and otherness: recognising the other/Other, and *becoming other*). This dimension is situated as *afterwards*, in the *afterwardsness*—a word translating Freud's '*Nachträglichkeit*' (in which we find the operation of retroactivation)—of speech; in that space so often left open by the analyst *as silence*. Drawing on an essay of Robert Fliess (Wilhelm Fliess' son, who himself became an analyst), Lacan fleshes out the position and operation of analytic silence in Seminar XII, stating: 'silence is the very locus where there appears the fabric on which there is unfolded the message of the subject[;] what is involved is precisely, at this level, its equivalence with a certain function of the [object *a*]', through confrontation with which, or refraction through which—as we know—the subject may find themselves, *afterwards, other than before*.[19]

Lacan's prescience here in revealing silence as the locus of the analytic—of the alterative—space (that very space of the cut) is demonstrated beautifully in an anecdote—an 'impromptu' collected by Jean Allouch—in which is recounted the enquiry a training analyst under Lacan's supervision made of him, in the words: '*but, Monsieur, this young man (the analysand she is consulting Lacan about) comes to me three or four times per week, tells me his stories to no end, pays me and goes. What is it that I have to give him in exchange?*' Lacan replied: '*Your silence!*'[20]

Notes

1. See Simon and Garfunkel, 'The Sounds of Silence', *Wednesday Morning, 3 A.M.* (Columbia Records: 1964, CL 2249).
2. Stanley Kubrick, in conversation with Maurice Rapf, quoted in *The Making of Kubrick's* 2001, p.102 (unnumbered page in photo insert).
3. James Gilbert, 'Auteur with a Capital A', in *Stanley Kubrick's 2001: A Space Odyssey: New Essays*, p.35.
4. Jacques Lacan, *The Seminar of Jacques Lacan, Book XI: The Four Fundamental Concepts of Psychoanalysis* [1964], ed. by Jacques-Alain Miller, trans. by Alan Sheridan (New York: W. W. Norton, 1981) p.26.
5. A couple of notable exceptions that Kubrick trusted were the films *Universe* (1960, dir. by Roman Kroitor and Colin Low; a documentary with innovative special effects, and from which Kubrick pinched Douglas Raine, who provided its narration), and the Cinerama extravaganza *To the Moon and Beyond* (which was shown at the 1964/1965 New York World's Fair; dir. by Lester Novros). *2001* was of course also presented in stunning Cinerama. Indeed, Kubrick's discernment was such that Clarke noted in his diary entry of December 10, 1964: 'Stanley calls after screening H.G. Wells's

Things to Come [dir. by William Cameron Menzies] and says he'll *never* see another movie I recommend.' See Clarke, *2001: A Space Odyssey*, pp.xxiv–xxv.

6. Christopher Frayling, *The 2001 File: Harry Lange and the Design of the Landmark Science Fiction Film* (London: Rare Art Press, 2015) p.234. For more on the scientific bases of the *Discovery*'s build and spaceflight—the technical coordination of which was mainly overseen by astronautic scientist Frederick I. Ordway and artist Harry Lange—see pp.210–303 (specifically on the evolution of the 'propulsion systems and fuselage details', pp.232–235), and its descriptions in Clarke's novel, the technologies underpinning which differ slightly to the film.
7. Piers Bizony, *2001: Filming the Future* (London: Aurum Press, 2000) p.40.
8. Philip Crosbie Morrison, quoted in *The Making of Kubrick's* 2001, p.51.
9. Ryle, *The Concept of Mind*, p.28.
10. Ibid.
11. Lacan, *Seminar XII: Crucial Problems for Psychoanalysis*, p.161.
12. Lacan, *Seminar X: Anxiety*, p.75.
13. Kojève, *Introduction to the Reading of Hegel*, pp.91–92.
14. Ibid., p.108.
15. Ibid.
16. Ibid.
17. For Kojève gives the definition: '*epistēmē* exists only in the cosmos in which there are ideas—i.e., eternal entities, having Eternity as their *topos*', in Ibid., p.113.
18. *The Seminar of Jacques Lacan, Book I: Freud's Papers on Technique, 1953–1954*, ed. by Jacques-Alain Miller, trans. by John Forrester (New York: W. W. Norton, 1988) p.107.
19. Lacan, *Seminar XII: Crucial Problems for Psychoanalysis*, p.161.
20. See Jean Allouch, *Les impromptus de Jacques Lacan: 543 bons mots recueillis* (Paris: Fayard, 2009) p.139. Quoted and translated by Owen Hewitson in the article 'Lessons from Lacan's Practice—Everyday Psychoanalysis, from the Classroom to the Boardroom (III)', available online: http://www.lacanonline.com/index/2014/02/lessons-from-lacans-practice-everyday-psychoanalysis-from-the-classroom-to-the-boardroom-iii/

CHAPTER 6

Surreality

Abstract In the concluding chapter, the surreality of the film's end is relayed first against Lacan's own Surrealist beginnings (his association with the art movement, and its proponents, and their influence on his psychoanalytic vision), and then put into conversation with Kojève's extrapolations from Hegel's work on desire, and its formation out of the Master/Slave dialectic. Bringing this back to the criteria of major and minor cuts, and departing from the cyclicity of the film and circularity of Hegel's science of Absolute Knowledge, the ending of the book attempts to offer up some of the ways made available by both *2001: A Space Odyssey* and Lacanian psychoanalytic theory of thinking differently politically and ecologically, and of making a radical cut into the fabric of our current conjuncture, and its underlying structurations.

Keywords Surrealism • Alexandre Kojève • G. W. F. Hegel • Desire • Politics • Feminism

Up to this point, we have surveyed a fair cross-section of Lacanian psychoanalytic concepts and tropes as they have arisen in the study of *2001: A Space Odyssey* that has been carried out over the preceding chapters (roughly designed to follow the trajectory of the motion picture). Theoretically, this has proceeded from our initial excavation of a theory of

D. Bristow, *2001: A Space Odyssey and Lacanian Psychoanalytic Theory*, The Palgrave Lacan Series,
https://doi.org/10.1007/978-3-319-69444-3_6

epistemic cuts—in accordance with the theorems outlined in Milner's condensations of Kojève and Koyré—in the 'Overture', through the quasi-allegorical conceptualisation of *das Ding* in relation to the film's enigmatic monolith, and a further honing of the theory, as well as a topologisation, of cuts in relation to the clinical Freudian structures, then to the wider hermeneutics of space and silence in the last two chapters. We now arrive at the surreality of the film's ending, which will be dealt with in relation to Lacan's early association with the Surrealist movement—and, initially, to a contemporaneous *écrit*: 'Beyond the "Reality Principle"'—but also through the lens of Kojève's philosophical postulations, arrived at through his reading of Hegel. That is; in the way in which Kojève describes Hegel's system, we too here come *full-circle*—back to Kojève's Hegel and its *science*—although, perhaps at the same time as recognising this, we should keep Deleuze's *difference* inherent in all repetition in mind, as much as Hegel's circularity (as uneasy bedfellows as these may generally be as systems).[1] It will be the aim of this chapter to work towards theorisations of both *work* and *desire*—and of the work *of* desire, and of what is, or can be, desired *beyond* work—and this through the attempts we will make at discovering what kinds of cuts might be excavatable, achievable, or realisable, in the work—or *working-through*—*of* this chapter.

'Beyond the "Reality Principle"' is a 1936 article as gripping as a modernist artistic manifesto, the thesis of which was originally intended by Lacan to have been completed by a sequel which never arrived. In this, it foreshadowed in some ways his originally planned eleventh Seminar, the topic for which was scheduled to be 'the Names-of-the-Father', which—after abandoning the course after its first session (at the host institution's bequest), and destroying all the remaining material—became a subject which Lacan vowed never to resurrect so sustainedly, because of the crushing ordination that forced him to relocate his Seminar, and re-theme it to the four fundamental concepts of psychoanalysis, in 1964.[2] The former, however, was simply never followed up as announced, but it left its readers on the tantalising tenterhook of Freud 'posit[ing] a "reality principle" whose role in his theory I propose to critique. But before doing so, I must first examine what has been provided by the studies that have been contributing to the new psychological science, alongside Freud's discipline, regarding *the reality of the image* and *forms of knowledge*.' Lacan ends by stating: 'these will constitute the two parts of my second article.'[3]

What is, or might be, the reality principle? In summary, Lacan characterises it in the abovementioned *écrit* as that which underwrites our subjective constitution, and which is made up from a constellation of images (Freud's *imagos*) that establish the *identificatory* (which, in Hegelian phrasing, might be re-termed the *identificatory moment*). Finding its groundwork in the order of the Imaginary, and its outpost in the ego, the *identificatory* is individually realised by the series of identifications that separate us from, and unite us to, each other, the *materialism* of which—to put a Kojèvean spin on it that is all the same already there in Lacan—is maintained by the process of *work* sustaining this subjective conjuncture.

In Surrealism, Lacan found an ally for his disparagement of the fragilely fortificatory imaginary systems of *identity* and *representation* on which the reality principle and the ego are based. As Thomas Brockelman puts it:

> Lacan, then, was fascinated by a cultural practice that seemed to demonstrate the possibility of breaking out of the "imaginary" identifications of the ego, that seemed to offer the promise of an alternative *ethos* to the "ego psychology" increasingly dominant in analysis. From Surrealism, Lacan could take evidence that the purpose of the psychoanalytic event was not to "put the ego in the place of the id" ([the standard reading of] "*Wo Es war soll Ich sein*" [*sic*]), but rather precisely to transgress the hold of the ego upon subjectivity.[4]

It wasn't only in Surrealism that the means to these ends could be found, but it was also of course right there in Freud, at whom Lacan arrived—at least in part of his understanding—through the Surrealists, for whom Freud was a key figure.[5] In Freud, it is the pleasure principle which disturbs—even *exposes*—whilst keeping in check, the images on which the reality principle is based and subsists. Lacan states that Freud had leant to the pleasure principle its psychoanalytic bearing by 'instating its signifying articulation of repetition in the circuit of reality' (as he put it in his introduction to the aforementioned article, in a 1966 addition to the *Écrits*).[6] That is, for subjectivity, and its jouissance, it is through and because of the pleasure principle that *reality*—and its *reality principle*—are manifested and maintained, reality *really* being that which skirts the impossible (the *Real*), by way of its (non-symmetrical) opposite, fantasy; the *possible*—in many ways—and the *only possible* (way) for the subject, constructed as it is by the Imaginary, in its interlacings with the Symbolic. (These theorisings of course go back to the distinction made at the beginning—in the 'Overture'—between the *Real* and *reality*.) Thus: how Miller paraphrases and diagrammatises Lacan's principle is as such:

> What Freud says, as heard by Lacan, is that the reality principle, is dominated by the pleasure principle, always [Miller inserts Fig. 6.1 here, as shown below].
>
> [...] The reality principle is the pursuit of the pleasure principle by other means—to parody Clausewitz.[7]

The surrealism of the Stargate sequence towards the end of *2001* is something like a shattering, a breaking-through, or going-beyond, of the image, which perhaps *enacts* the pleasure principle, or indeed rather goes *beyond* it (as both Freud and Kubrick indicate in their respective titular uses of the word 'beyond'), as if the barriers that constrain 'RP' (the reality principle) in Miller's schema (Fig. 6.1) have *dissolved*—in an instance of evisceration of a *screen*, or the *cut(ing out) of a cut*—opening out, *beyond* pure 'PP' (pleasure principle); that is, into the *lack of distinction* between 'RP' and 'PP', or, in extension, between any two unificatory/separatory 'sides'. In *2001*, this occurrence is what we term here a 'surreality'; that is, a (*surreal*) *reality* in the filmic universe of which it is a part, a *reality* for the character—Dave Bowman—experiencing it (whilst 'surreal*ism*' implies an intentionality behind that which has been made surreal). Before encountering the Stargate's surrealistic surreality in itself, we will defer to R. Barton Palmer's summary of the challenges to cinematic conventionality that *2001* presented, in general: (As a movie for which the second promotional poster ran the tagline: 'the ultimate trip'.)

> It is hard to imagine, in fact, how Kubrick could have further undermined the ideological presuppositions of Hollywood storytelling, with its narrative driven by and centred on character, its embodiment of political or social questions in valued individual destinies capable of neat resolution, its conventional deployment of admirable protagonists and despicable antagonists, its devotion to arousing the sympathy and pathos necessary for the proverbial happy ending, and, perhaps most important, its confection of a closure that masked contradiction with a perfect knowledge of character and event. Such a sweeping rejection of those structures and themes so customary in the American commercial cinema certainly challenged the analytical and critical acumen of journalists and reviewers at the time.[8]

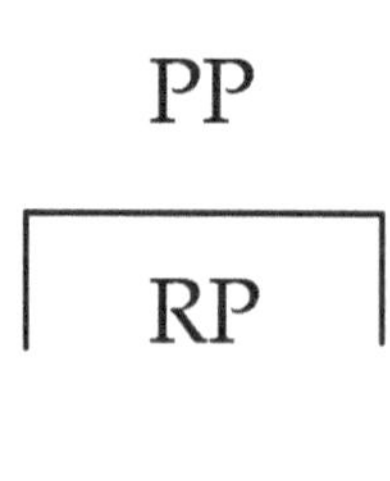

Fig. 6.1 Reality principle under the domination of the pleasure principle. Source: the author

Beyond *2001*'s structure and plot, which in themselves reject the linearities of the Hollywood standard, is the *avant-garde*-invoking last act of the film. Utilising innovative special effects, the Stargate sequence is prefaced by a wonderful panorama of Jupiter and its moons—set to the rising intensity of Ligeti's *Lux Aeterna*—into which enters the monolith, floating in tandem with these celestial bodies, as well as the sun, *Discovery*, and Bowman's spacepod, which we see exit the ship on its last EVA. When the planet, moons, and monolith are all aligned in their orbits, from this vista the camera spectacularly pans upwards to establish a steady shot amongst the deep black backdrop of space punctured by pinholes of light that are the distant stars. Then at the screen shoots a flash of fuzzy red as the stars fade and we enter their beyond to an array of multicoloured and multi-shaped visuals, through which we seem to pass, towards the vanishing point of a slit of pure black at their centre, which splits them seemingly into two walls, non-symmetrical to—but simultaneous with—each other. This feat of imagery was achieved by the employment of slit-scan photography, the modern adaptation and operation of which was overseen by special effects supervisor Douglas Trumbull, and it is used to tremendous effect, producing a spectrum of patterned colour and vortices of light. (Trumbull's adapted technique was influential, and inspired effects in the title sequences of the 1970s' *Doctor Who* series; *Star Trek: The Next Generation*'s warp speed depictions; and the final scenes of Christopher Nolan's *Interstellar* (2014) in the tesseract.) The light and colour are initially blocked into lines recalling at moments the jutting angularity of Vorticism and at others the ordered rigidity of De Stijl, before themselves setting the phosphorescent lightshow standard for the look of trippy *spaciness*, jumping between verticality and horizontality; all this whilst cutting back to Bowman's face, at first seemingly undergoing the strains of some kind of force without g, so to speak—as there are no signs of the folds in the skin of the cheeks characteristic of g-force, implying something beyond *acceleration*—before the startling cuts to his frozen contorted visage, as already discussed. Almost as if we are being taken on a journey through something like *electricity*—the camera-eye's view of transport through electrical or telephone cables is often depicted similarly, such as at the beginning of Krzysztof Kieślowski's *Trois couleurs: Rouge* (1994)—we are met with an onrush of light before the cut to the colourised eye blinking blue and yellow. From this image, we re-enter the Stargate's surreality at slower-moving manifestations of light that could be a gazing-at-galaxies *à la* the aurora borealis, or strange supernovae, or deeply outer-space phenomena (perhaps closer in the universe's space-time to the Big Bang). The milkiness of the images that follow was brought into effect by experimentation with coloured liquids in paint trays, and it furnishes an uncanny effect of terrestriality/extraterrestriality at once: images of mammoth unknowable galactic occurrences mirroring the mundanity of spiders' webs; stretching facelike globules; and even the entry of a sperm, with its tail, into an ovum. These scenes are choreographed to the changing orchestration of the chromatic sections of repose in Ligeti's *Atmosphères*, which soundtrack the sequence, and are then cut into by another negativised eye, highlighted in fluorescent blue and purple, which brings in the hovering diamonds, set against a static display of stars, and carpeted by slit-scan colours beneath (as alluded to above). After the next eye, orange and blue, is a milky luminescence of a kind of firmament encountered from below, somehow sea- and cloud-like at once, which intersperses with another eye-shot, purpler and bluer, which beckons the alien landscapes that are travelled over, again worldly and otherworldly: rocky expanses, gorges, volcanic-looking deeps, seas—seemingly reflective, but of different colours and images to those above them—and a sky of gaseous combinations beyond comprehension. Floating over these areas becomes tranquil and meditative, especially in synch with the score, until at a lull in sound and glide the eye is returned to, this time blinking through colour combinations until Bowman's hazel iris reappears, and is lingered on.

Thus we find ourselves in 'the room at the end', looking out of the spacepod window, at a scene that is slightly Magrittean in its strangeness-*rendered-normally*. With the clanging noises and modulated voices of Ligeti's altered *Aventures*, we are party to exterior shots of the spacepod, from inside the room, whilst the sound of muffled breathing returns. Bowman is then in the interior of the room (there is no implication of an exterior), in his spacesuit, and his craft has vanished; he has aged considerably and encounters a mirror-stage in the bathroom confirming this strange fragmentation, before seeing himself thereafter at older stages—in provided (bed)clothes, and without breathing apparatus—the return of the gaze of which is left as a missed encounter each time, by the then-absence of the former self. This *mise-en-scène* thus progresses to the final transformation before the monolith, and the ascent of the enwombed, wide-eyed foetus, the gaze of whom turns to the world before the final cut to black and the end of the film.[9]

The surrealism at the end of *2001* leaves us placed, and left, on the cusp of a last overturning—of *subjectivity*, but perhaps also of *objectivity* itself;

even of *universality* (if it is to be the cusp of a hegemonisation)—the lead-up to which, through the film's major and minor cuts, it might now be worth recapitulating, perhaps in the *historical materialist* mode as conceptualised by Walter Benjamin, for whom: 'the historical materialist blasts the epoch [for us, the *epistēmē*] out of its reified "historical continuity," [...] result[ing] in the simultaneous preservation and sublation [*Aufhebung*] of the lifework *in* the work, of the epoch *in* the lifework, and of the course of history *in* the epoch.'[10] It will be through Kojève that we will track how these sublations come to be, through the epochs—and their underlying epistemic shifts—of Kubrick and Clarke's *Space Odyssey*.

In a contentious note that Kojève added to his *Introduction to the Reading of Hegel* in 1946, he outlines what he terms 'the disappearance of Man at the end of History' as he then envisaged it in Hegel and Marx; he later comes to heavily revise this reading in a 1948 addendum to the note, whilst preserving the original before—and thus then, *sublatively*, *alongside*—it. In this note, he stresses that:

> The end of human Time or History [...] means quite simply the cessation of Action in the full sense of the term. Practically, this means: the disappearance of wars and bloody revolutions[.] All the rest can be preserved indefinitely; art, love, play, etc., etc.; in short, everything that makes Man *happy*. Let us recall that this Hegelian theme, among many others, was taken up by Marx. History properly so-called, in which men ("classes") fight among themselves for recognition and fight against Nature by work, is called in Marx "Realm of necessity" (*Reich der Notwendigkeit*); *beyond* (*jenseits*) is situated the "Realm of freedom" (*Reich der Freiheit*), in which men (mutually recognizing one another without reservation) no longer fight, and work as little as possible[.] Cf. *Das Kapital*, Book III, Chapter 48, end of the second paragraph of § III.[11]

The ambiguity of this note (made pronounced by its lengthy revision) chimes with the ambiguity of the end of *2001*, and what might be, or become of, its final transformation. However, we will here engage with those elements which Kojève identifies as spurring the (dialectical) *becoming* of History. They are, in paraphrase: the *Fight* (and *fights*) and *Work* (and *works*).

Hegel's Master/Slave dialectic is based on a fight *to the death* for pure prestige, or *recognition* (by the other), in which neither party end up dying, as one—who fearing death, chooses not to die (a quite sensible decision)—comes to assume the role of Slave, promoting the other to the role of Master, for whom the Slave will then come to work (but in the knowledge that the Master will one day die, and things will be different); a decision the Slave thus makes *consciously* and *freely*, as Kojève insists.[12] As Kojève draws out perspicaciously, what is key to the fight is *Desire*, the Hegelian—and subsequently Lacanian—conceptualisation of which Kojève gives a masterful commentary on at the beginning of his great exegesis, which takes the form of a heavily annotated presentation of section A of Chap. 4 of the *Phenomenology of Spirit*, 'in place of an introduction'. In this, he summarises:

> Anthropogenetic Desire is different from animal Desire [...] in that it is directed, not toward a real, "positive," given object, but toward another Desire[.] Desire is human only if the one desires, not the body, but the Desire of the other; if he wants to "possess" or to "assimilate" the Desire taken as Desire—that is to say, if he wants to be "desired" or "loved," or, rather, "recognized" in his [...] reality as a human individual. Likewise, Desire directed toward a natural object is human only to the extent that it is "mediated" by the Desire of another directed toward the same object: it is human to desire what others desire, because they desire it. Thus, an object perfectly useless from the biological point of view (such as a medal, or the enemy's flag) can be desired because it is the object of other desires.[13]

Mapping this onto the Master/Slave dialectic, Kojève points out that:

> On the one hand, the Master is Master only because his Desire was directed not toward a thing, but toward another desire—thus, it was a desire for recognition. On the other, when he has consequently become Master, it is as Master that he must desire to be recognized; and he can be recognized as such only by making the Other his Slave[.] Thus, finally, his Desire is directed toward a thing, and not—as it seemed at first—toward a (human) Desire[.] After the fight that made him a Master, he is not what he wanted to be in starting that fight: a man recognized by another man. Therefore: if man can be satisfied only by recognition, the man who behaves as a Master will never be satisfied. [Thus,] the satisfied man will necessarily be a Slave; or more exactly, the man who has been a Slave, who has passed through Slavery, who has "dialectically overcome" his slavery.[14]

Before the satisfaction of the Slave, however, comes his *Work*. This work is in the service of the Master, but it goes further, in encompassing something that the Master cannot do; that is, extend mastery over *Nature* (this is thus, in effect, a sublation—of the Slave's—of the Master). It is the

Slave that masters Nature, through work; that is, through fashioning nature into the best realisation of that which is demanded of it (a demand albeit made by the Master, but only realisable through the work of the Slave). Whilst the *Fight* gives a parabolic physical act to the *place of the cut* that brings in this dialectic, we can see the beginning of such a relation in *2001* in the monolith's inauguration of the work of war, in the *technê* of the first weaponisation (of the bone). Indeed—contrary to the declaration that Clarke makes when Moon-Watcher picks up the bone: that now he's 'master of the world'—the apes, in their epistemological segue, have become enslaved again (whilst being emancipated of their environmental blindness); this time by the work that will be demanded of them in the service of the war relation for the next several million years (with all the advancements that will be made within it considered).[15] Kojève says as much himself:

> It is not the Fight as such, the risk of life, but *Work* that one day produces a machine gun, and no longer an ax. The purely warlike attitude of the Master does not vary throughout the centuries, and therefore it cannot engender a historical change. Without the Slave's Work, the "first" Fight would be reproduced indefinitely: nothing would change in it[.]
>
> Quite different is the situation created by Work. Man who works *transforms* given Nature[.] After making the first ax, man can use it to make a second one, which, by that very fact, will be another, a better ax. Production transforms the means of production[.] Where there is Work, then, there is necessarily change, progress, historical evolution.[16]

Historical evolution (what we might call history's series of minor cuts: adaptive, improving, *revolutionary*, in the sense of Lacan's Copernicus)—within a historical constellation, a historical continuity—thus occurs, and, by the Slave's work, expands the technological potential of its prescription, and even perhaps paves the way to a next epistemological shift, creating the conditions for genuine change (in contrast to which, the monolithic major cut must of necessity be reconceptualised as something like the intervention of a 'divine violence', to use Benjamin's term). The dialectic of this work here established, though, has to take into account something that Kubrick was acutely aware of, if not in so many words, and as has been previously alluded to; that is, whether the (war) relation is determined by the *technê*, or has control over it. To bring this into the specific discussion, in its elucidation: when a civilisation becomes capable of manufacturing weapons of mass destruction and devastation, the question arises as to whether we are at the mercy of their sublime power (and the instabilities that could unleash it), or if they are subordinate to our mastery and responsibility. Indeed, Jeremy Bernstein broached this question, in a write-up on Kubrick, in relation not only to human interaction with said power, but extraterrestrial too, in which he centralises the awesome force of that power itself:

> Any civilization that operates on the same laws of nature as our own will inevitably reach the point where it learns to explore space and to use nuclear energy about simultaneously. The question is whether there can exist any society with enough maturity to peacefully use the latter to perform the former. In fact, some of the more melancholy thinkers on this subject have come to the conclusion that the earth has never been visited by beings from outer space because no civilization has been able to survive its own technology.[17]

Lacan of course was also attuned to the utter powerlessness that this power, and its cultivation, threatens, and is able, to render. As we have already seen, he locates his own nuclear master/slave dialectic in the field of science, and its guided teleologies. For Kubrick, as we know, it was something of an obsession. Expanding on Bernstein, in an interview on *2001*, he declaimed:

> Unhappily, the infant-mortality rate among emerging civilizations in the cosmos must be very high. Not that it will matter except to us; the destruction of this planet would have no significance on a cosmic scale; to an observer in the Andromeda nebulae, the sign of our extinction would be no more than a match flaring for a second in the heavens; and if that match does blaze in the darkness, there will be none to mourn a race that used a power that could have lit a beacon in the stars to light its funeral pyre. The choice is ours.[18]

The ultimate characterisation of the Cold War is that of a stalemate, for—if thawed—beyond it lies what I have called elsewhere an unknown known—that is, *unconsciously disavowed*—radical contentlessness, or, as it used to be called: mutually assured destruction.[19] Whilst *2001* might dare to think otherwise to this result—nuclear power being deployed technologically in the propulsion systems of ships that can thus reach far-off planets such as Jupiter—its *technê* is still within the stalemate of the work of the war machine (nuclear weapons are poised in Earth's orbit also, no doubt with litany of programmed coordinates and targets). Clarke, in the book, gives an indication of where the nuclear theme might go from the final cut, with the Star Child blowing up the orbiting weapons, which—although he confirms this as a repletion of the Star Child's energy reserves in *2010* (which, beyond being a revolutionary immunity, becomes even a form of sustenance, thus inaugurating *what could be* a truly new dimension of being, although it is not exploited as such in the follow-ups)—as we have seen, is still ambiguous, in and of itself. Nonetheless, the end of *2001* leaves us with a radical contentlessness too, but one indicating the possibility (at least) of a *productive* beyond, rather than an apocalyptic one.

Indeed, its final transformative cut is a postulation beyond the Hegelian schema, in which knowledge—having completed itself—restarts at the beginning of the circle, but with the sublative quality absent from Nietzsche's *eternal return of the same*—in which *exact recurrence* is given a hysterical affirmation—that always also haunts that final image, and is of course interlaced with the film's bank of references.[20] It's a cut that cuts through the circle—as the major cuts that preceded it had through theirs (from that of animality to humanity in the first, from artificial intelligence to computer sentience in the second, and from the humanly-conceived space-time of objectivity to that of the Stargate in the third)—but in this it remains (to us) an imaginary cut, an unrealisable (by us) cut of monolithic providence (or 'divine violence'; perhaps that of the '*avenir*', or 'to-come', in Derrida's messianic terminology). Thus, with this cut—which opens up a rainbow-coloured portal of possibility—as inspiration, we can redirect this treatment along the Lacanian (and Hegelo-Kojèvean), psychoanalytic route of a work—a *working-through*—that can produce its own cut; a change in the texture of the weave as it stands.

Kojève indeed states towards the end of his text that 'the only way in which one can "go beyond" given-Being without sinking into pure Nothingness [is] namely in and by thought.'[21] From this, we can attempt to posit what might thus be the *thinkable* cut, which, through thought—and its work—might beckon what we could be able to bring about, sublatively.

In the operation of *desire*, a *beyond* is posited, towards which the subject may work (although they might need a push. Indeed, desire is the main motivator for such work, but it can come to be thwarted by the machinations of the unconscious, manifesting in symptoms, slips and bungled actions, obsessional detours, actings out and *passages à l'acte*, etc.). Not ceding on this—potentially truly transformative—desire, which in Lacanian psychoanalysis is perhaps the main obstacle that is come up against in its analysands, Lacan bills: 'the ethics of psychoanalysis.'[22] We thus might here theorise the *ethical* as the place of (the possibility of) the major cut—that which will *separate a system of thought of Antiquity from a modern system of thought* (to remind ourselves of Milner's Kojève)—and lend to it the definition that Kojève gives in his *Introduction to the Reading of Hegel* to the *progressive*: 'I do not believe that progress can be defined otherwise than in the following manner: there is progress from A to B, if A can be understood from B but B cannot be understood from A.'[23] This is precisely the formula of the major cut. In the film, the B-understanding has always to be effected by the monolith: the A-understanding would not have got to the B-understanding by itself. This renders the cuts that are inaugurated *major*. Yet—as discussed above in relation to *work* (for example, on nuclear power)—it might be(come so) that the paving towards a major cut can be laid without *such stones* (or their second—or however-many—coming, the waiting for which could then be shown up as a deferral, or procrastination, of action, or of agency: *the absence of the work*, as Michel Foucault once said).

To combine Hegel and Heidegger once again—as we had at the beginning—we might here reread (or again *overdetermine*) the concision '*Spirit is a bone*' as implying that the bone (the specific transformative bone raised aloft by the ape) occupies the place of *animator* of the *Spirit*—or the agency—of *Being-in-the-World*; thus awakening realisation to it, but also be(com)ing part of it itself; this, in the dialectical movement encapsulated by Kojève in reference to the *transformation of the World by the one whom that World forms.* For Kojève, the one who occupies this position is the Slave:

> Only the Slave can transcend the given World (which is subjugated by the Master) and not perish. Only the Slave can transform the World that forms him and fixes him in slavery and create a World that he has formed in which he will be free. And the Slave achieves this only through forced and terrified work carried out in the Master's service. To be sure, this work by itself does not free him. But in transforming the World by this work, the Slave transforms himself, too, and thus creates the new objective conditions that permit him to take up once more the liberating Fight for recognition that he refused in the beginning for fear of death. And thus in the long run, all slavish work realizes not the Master's will, but the will—at first unconscious—of the Slave, who—finally—succeeds where the Master—necessarily—fails.[24]

Integral to the modernist art movements—of which Surrealism reigned as one of the most supreme in this respect—was the *manifesto* (André Breton was unwavering about adherence to his own for the Surrealist movement). Indeed, what *could*, even *might*, *become manifest* from Kojève's words implies the *hope-desire*—the 'at first unconscious' *will*—that is the motor of the conception, or postulation, of a *beyond-reality* (as *surreality* itself is open to being rephrased).

It is for the intersectional constellation of current struggles (structural, political, cultural, racial, sexual, social, ecological, and so many more)—those through which there is hope for the disappearance of war and *bloody* revolutions, as Kojève says (the possibility of which would have pleased Kubrick and Clarke); hope for a yet-to-arrive *justice*; hope, that is, for a *major cut*—that this motor needs to be kept fuelled; and this by the work that might bring about *new dimensions*: a revolutionary epistemological overhauling of the (current) *epistēmē*. This might be one realisation, or *interpretation*, at which *2001* delivers us. Thus, as its audience, it is perhaps imperative to look for extensions *beyond* its (cyclical) filmic universe, which is of course the universe at large. To do so might indeed constitute a form of *cut* (even *major*, in the sense of making a departure from the film's monolithic (non-circular) cyclicality). We can perhaps do so *temporally*, which is to say: we might look beyond 1968, that revolutionary year, and look too beyond 2001, that reactionary year (in which nothing like the space age envisioned in *2001* had really come about) to our *now*, and try to find resonances. One such might be found in the yoking of two of the abovementioned movements (which work towards transforming the World and thus towards transforming ourselves through doing so) made by Juliet Mitchell in her essential, forward-thinking work *Psychoanalysis and Feminism*, in which she lays out some fundamental considerations concerning the practicabilities of *cuts* (*in*, or *to*, consciousness; that is, here: *in* feminist and working-class consciousness, and *to* patriarchal and capitalist consciousness):

> Because patriarchy is by no means identical with capitalism, the successes and strengths of the two revolutionary movements will not follow along neatly parallel lines. [...] There is no question of either political movement taking precedence, or of either revolutionary group being mutually exclusive or even of each group containing only its denominational membership. By this I mean that just when the working class becomes revolutionary, people who do not actually come from the working class can make a political

> transformation of their own class origins and join it, so *when* the feminist movement has a revolutionary theory and practice, men too (if with difficulty) can give up their patriarchal privileges and become feminists.[25]

In Kojève's terms, this instance perhaps implies a sublative moment—or a point of *vanishing mediation*—in which, in a flash, B becomes somewhat understandable, possible, attainable to A, and after which A will be forever *gone beyond*. As Mitchell states poignantly: 'this is not to say that [such men] can become members of the movement where it operates at the level of feminist consciousness any more than Marxist intellectuals can join the trade union movement which is the equivalent of working-class consciousness—they can merely support it in a practical fashion', but this, perhaps, in that Marxian 'mutual recognition of one another without reservation' that Kojève envisioned at his most idealistic.[26] Thus, it is through such *transformed-transforming* consciousness, and its work (and the work intersecting with, and in support of, it) that a *beyond* can be postulated, and aimed at; that a *modernity* can be established, establishing its *before* as *Antiquity*; that, in this example, the Dusk of *Man*—as *2001* implies by its first subtitle—can be brought about. It will be left to *desire* to cut through that which (*microfascistically*) seeks to compromise it, and to slowly lift the repressions of the unconscious, and create it anew: *freely*, *consciously* (as it is in Kojève's Hegelian aspiration). For as Lacan said, in relation to the 'R' schema: 'the barred S of desire, props up the field of reality'; to *assume*—and to *not give ground relative to*—this desire can thus modulate that field of reality's (or the *field of vision*'s, and *field of visions*') *principle*.

As Kubrick frames it (with, and throughout, *2001*): the choice is ours.

Notes

1. Kojève describes 'Hegel finally com[ing] to a point that is none other than his point of departure: the *final* synthesis is also the *initial* thesis. Thus he establishes that he has gone around or described a *circle*, and that if he wants to continue, he can only *go around again*: it is impossible to *extend* his description; one can only *make it again* as it has already been made once.

   This means that Hegel's discourse exhausts *all* the *possibilities* of thought.' See Kojève, *Introduction to the Reading of Hegel*, p.194. The Deleuzian point is to realise that all repetition implies difference even if only that of the repetition not being its first time, *qua* repetition.

2. As Miller explains: 'dramatically interrupted by Lacan's demotion from the rank of *didacticien* (which at the time meant a psychoanalyst authorized to train other psychoanalysts [i.e., a "training analyst"]), his Seminar began anew in January 1964 in the rue d'Ulm at the École Normale Supérieure with a new title: "The Four Fundamental Concepts of Psychoanalysis."

   Lacan always refused to revisit the theme of the Seminar that was cut short and even to publish while he was alive the text of the single class he had given. Having concluded from his tribulations that "psychoanalytic discourse" had not authorized him to lift, as he had intended, the veil Freud had cast over the true mainspring of psychoanalysis, and that he had been struck down for his sacrilegious act, he signaled, in a word to the wise, in particular in the ironic title that he gave a later Seminar, *Les non-dupes errent*, that he would keep close to his chest truths that were too tempestuous.' See Jacques-Alain Miller, 'Foreword' [2005], in *On the Names-of-the-Father*, trans. by Bruce Fink (Cambridge: Polity, 2013) p.vii.
3. Lacan, 'Beyond the "Reality Principle"', in *Écrits*, p.75.
4. Thomas Brockelman, 'Lacan and Modernism: Representation and its Vicissitudes', in *Disseminating Lacan*, ed. by David Pettigrew and François Raffoul (New York: State University of New York Press, 1996) p.208.
5. Indeed, Roudinesco points out in her biography of Lacan that much of Freud's reputation in France was being fostered by interpretations of his work being given by 'the literary and philosophical avant-gardes', and identifies that it was in 'about 1923 [that Lacan] heard about Freud's theories for the first time'; that is, at a key time in which he 'was also interested in dadaism and soon discovered the new outlook and early manifestations of surrealism through the review *Littérature*. He met André Breton and Philippe Soupault and listened spellbound, at Shakespeare & Co., to the first readings of James Joyce's *Ulysses*.' See Roudinesco, *Jacques Lacan*, p.15 and p.13.
6. Jacques Lacan, 'On My Antecedents' [1966], in *Écrits*, p.53.
7. Jacque-Alain Miller, 'Everyone is Mad' [2008], trans. by Adrian Price, in *Culture/Clinic 1: Applied Lacanian Psychoanalysis*, ed. by Jacques-Alain Miller and Maire Jaanus (Minneapolis: University of Minnesota Press, 2013) p.34. Carl von Clausewitz famously wrote that '*war is the continuation of policy with other means.*' See Carl von Clausewitz, *On War* [1832], trans. by Michael Howard and Peter Paret, abridged by Beatrice Heuser (Oxford: Oxford World's Classics, 2007) p.7.
8. R. Barton Palmer, '*2001*: The Critical Reception and the Generation Gap', in *Stanley Kubrick's 2001: A Space Odyssey: New Essays*, p.16. In many ways *2001* abides more by Maurice Merleau-Ponty's idea of the cinematic work-of-art: 'cinema has yet to provide us with many films that are works of art from start to finish: its infatuation with stars, the sensationalism of the

zoom, the twists and turns of plot and the intrusion of pretty pictures and witty dialogue, are all tempting pitfalls for films which chase success and, in so doing, eschew properly cinematic means of expression. [...] Beauty, when it manifests itself in cinematography, lies not in the story itself, which could quite easily be recounted in prose and still less in the ideas which this story may evoke[.] What matters is the selection of episodes to be represented and, in each one, the choice of shots that will be featured in the film, the length of time allotted to these elements, the order in which they are to be presented, the sound or words with which they are or are not to be accompanied. Taken together, all these factors contribute to form a particular overall cinematographical rhythm.' See Maurice Merleau-Ponty, *The World of Perception* [1948], trans. by Oliver Davis (Abingdon: Routledge Classic, 2008) p.73.

9. Clarke's wonderful diary entries towards the end of 1964 almost seem like manifesto points themselves, for the ending—indeed, the entirety—of the movie: 'October 17. Stanley has invented the wild idea of slightly fag robots who create a Victorian environment to put our heroes at ease.

   November 28. Phoned Isaac Asimov to discuss the biochemistry of turning vegetarians into carnivores.

   December 10. Stanley calls after screening H.G. Wells's *Things to Come* and says he'll *never* see another movie I recommend.

   December 24. Slowly tinkering with the final pages, so I can have them as a Christmas present for Stanley. Clarke, *2001: A Space Odyssey*, pp.xxiv–xxv.
10. Walter Benjamin, 'Eduard Fuchs, Collector and Historian' [1937], trans. by Howard Eiland and Michael W. Jennings, in *The Work of Art in the Age of its Technological Reproducibility and Other Writings on Media*, ed. by Michael W. Jennings and others (Cambridge, MA: The Belknap Press of Harvard University Press, 2008) p.118.
11. Kojève, *Introduction to the Reading of Hegel*, p.159, n.6.
12. See Ibid., p.65. He here makes the vital distinction between the Slave and the Bourgeois, who does not *freely and consciously* choose, in that: 'the Bourgeois' problem seems insoluble: he must work for *another* and can work only for *himself*. Now in fact, Man manages to resolve this problem, and he resolves it once more by the bourgeois principle of private *Property*. The Bourgeois does not work for another. But he does not work for himself, taken as a biological entity, either. He works for himself taken as a "legal *person*," as a private *Property-owner*: he works for Property taken as such—i.e., Property that has now become *money*; he works for Capital.

    In other words, the bourgeois Worker presupposes—and conditions—an *Entsagung*, an *Abnegation* of human existence. Man transcends himself, surpasses himself, projects himself far away from himself by projecting himself onto the idea of private Property, of Capital, which—while being the Property-owner's own product—becomes independent of him and enslaves him just as the Master enslaved the Slave; with this difference, however, that the enslavement is not conscious and freely accepted by the Worker. (We see, by the way, that for Hegel, as for Marx, the central phenomenon of the bourgeois World is not the enslavement of the working man, of the *poor* bourgeois, by the rich bourgeois, but the enslavement of *both* by Capital.)'

13. Ibid., p.6.
14. Ibid., pp.19–20.
15. Clarke states in the novel that 'now [Moon-Watcher] was master of the world, and he was not quite sure what to do next', and had written the same in the screenplay. See Clarke, *2001: A Space Odyssey*, p.33, and Kubrick and Clarke, *2001: A Space Odyssey, Screenplay*, p.a26, 10/13/65.
16. Kojève, *Introduction to the Reading of Hegel*, p.51.
17. Jeremy Bernstein, 'Profile: Stanley Kubrick' [1966], in *The Making of Kubrick's* 2001, pp.59–60.
18. Kubrick, 'Stanley Kubrick: *Playboy* Interview', in *The Making of* 2001: A Space Odyssey, p.300. In the same interview, Kubrick offers a religio-philosophical analysis of the phenomenon of Cold War desensitivity, which gets to the heart of the death drive (as opposed to 'death wish', as Kubrick points out): 'as fewer and fewer people find solace in religion as a buffer between themselves and the terminal moment, I actually believe that they unconsciously derive a kind of perverse solace from the idea that in the event of nuclear war, the world dies with them. God is dead, but the bomb endures; thus, they are no longer alone in the terrible vulnerability of their mortality. Sartre once wrote that if there was one thing you could tell a man about to be executed that would make him happy, it was that a comet would strike the earth the next day and destroy every living human being. This is not so much a collective death wish or self-destructive urge as a reflection of the awesome and agonizing loneliness of death. This is extremely pernicious, of course, because it aborts the kind of fury and indignation that should galvanize the world into defusing a situation where a few political leaders on both sides are seriously prepared to incinerate millions of people out of some misguided sense of national interest' (p.295). In a beautiful and terrifying montage very reminiscent of Kubrick—in both image and sound, which, respectively, seem influenced by the Stargate sequence, and *The Shining's* score—David Lynch focuses minutely on nuclear detonation, and gives it an incredibly eerie originarity, in the *Twin Peaks* episode, 'Gotta Light?' (series 3, 2017).
19. See 'The Campaign for Nuclear War and the Unknown Known', in EDA Collective, *Politactics: Political Conversations from Everyday Analysis* (Alresford: Zero Books, 2016) pp.71–74.

20. To one early viewer, the film's circularity was completely transparent, as is recounted in a letter to Kubrick from M-G-M publicist Edward Lawrence, concerning his wife's response to the movie: 'in discussing the philosophy of the film, she kept repeating, "The beginning is the end, and the end is the beginning."' See Edward Lawrence, quoted in *The Making of Kubrick's* 2001, p.184.
21. Kojève, *Introduction to the Reading of Hegel*, p.254.
22. See Lacan, *Seminar VII: The Ethics of Psychoanalysis*, p.392.
23. Kojève, *Introduction to the Reading of Hegel*, p.87, n.3. This thus represents the *dialectic of the beyond*. For example, in the idea (previously given) of a *beyond of knowledge*, what this will be—its *truth*—cannot be known (cannot be acknowledged), although an 'idea' of it can be had. Thus, *knowledge* is at position A, and its *beyond* at position B, but how B will come to 'know'—or 'understand'—A also cannot be known to A.
24. Ibid., pp.29–30.
25. Juliet Mitchell, *Psychoanalysis and Feminism* [1974] (London: Penguin Books, 1990) pp.414–415.
26. Ibid., p.415.

Index[1]

[1]Note: Page number followed by 'n' refers to notes.

D. Bristow, *2001: A Space Odyssey and Lacanian Psychoanalytic Theory*, The Palgrave Lacan Series,
https://doi.org/10.1007/978-3-319-69444-3

L

Kolibri 360 GmbH
Ferdinand-Jühlke-Straße 7
99095 Erfurt, Deutschland
produktsicherheit@kolibri360.de